# CONSTRUCTION OF RANDOM SIGNALS

# FTROM THEIR HIGHER ORDER MOMENTS

by

I0474039

ISMAIL CHAMSEDL

Thesis submitted for the degree of

Doctor of Philosophy of the university of London

and the Diploma of Membership of Imperial College

May 1997

Department of Electrical and Electronic Engineering

Imperial College of Science, Technology and Medicine

University of London

# Abstract

The problem of estimating a probability density function from its higher order moments specified up to a certain finite order is formulated and discussed in a unified framework for higher order statistics. Two approaches are identified: The least squares approach and the maximum entropy approach. These are shown to result from deeper considerations of the concepts of complexity and duality, and are discussed thoroughly together with other possible methods, namely the empirical distribution function and the parameter estimation approach, and complete solutions are provided. A number of simulations shed light on the problem. The various methods are then compared and the result is a unified solution for the two proposed methods.

# Statement of originality

The following results are believed to be novel and original contributions presented in this thesis:

1. Formulation of the general problem of estimation of a probability density from a finite prespecified set of its moments, and its interpretation in terms of complexity and duality.

2. The review of the empirical distribution function method and its link with the moments approach, and the interpretation of the method of parameter estimation of a probability density in the context of the weighting function.

3. Formulation of the least squares estimation method in terms of the weighting function, and solution of the problem.

4. Formulation of the maximum entropy method and solution of the problem.

5. Derivation of the equations linking moments and coefficients of the solution probability density corresponding to the maximum entropy method.

6. Complete solution, unifying the two proposed methods (items 3 and 5), showing their equivalence.

# Table of Contents

# List of figures

# List of Tables

# CHAPTER 1

# INTRODUCTION

Signal processing is about extracting information from observations. This statement is sufficiently general and operational as long as the scope of the words "information" and "observations" is adequately specified, for these two concepts are subjective to a certain degree: The information required may be the spectral composition of a time series, the internal description of the system according to its observed output, or the values of a signal outside the range of observation (limited time window or bounded set in physical 3-dimensional space, for example). On the other hand, an "observation" can be anything measurable, in the physical sense. However, it is important to note here that an observation may be in itself a signal processing task, and therefore can be classified as information, in the previous meaning, but corresponding to a subprocess at a lower level in a hierarchy: A tachometer measuring the rotational speed must process its output voltage to clean it from pulses due to commutations, an eye reading the indication of a voltmeter or ammeter will have an error due to its limited resolution, while the device itself, if electromechanical, will indicate only the mean value of relatively rapid fluctuation voltage. We conclude from this remark that there's a kind of duality between these two concepts of information and observation. To be able to start inspection of a system, an additional statement, the role of which is to merge the two concepts at the starting point, is needed: The observations (or information) must be coded in a certain **alphabet** . In practice, this would be the real numbers or integers (equivalent in this context), a logical variable (TRUE or FALSE), or the syntax of a programming language.

## 1.1. Formulation of the problem

Having settled this point, we can say that, in this view, signal processing is the way the human mind interacts with and responds to the outside world, and therefore it is a basic and fundamental human activity. Thus we could expect that all the different problems in signal processing can be cast in a unified framework, which will represent, in technical and scientific terms, the common way the human mind pursues in its interaction with the physical world. In fact, such a unified view is needed because of the multitude of estimation methods used to process the data in hand and, more importantly, the assumptions made regarding these data for the purpose of simplification of the problems encountered. In general, there's no basis for these assumptions other than the analytical tractability of the problem so formulated or the agreement between theoretical results and **some** specific experiments (simulations, tests on models, etc.). For example, the periodogram method used in spectral analysis assumes a zero value for the observed signal , like $(x_k)_{0 \leq k \leq n}$, outside the time observation window i.e.

$$x_k = 0, k \geq n+1$$

an assumption which is not always valid. Another example, also in spectral analysis is in  the parametric methods where a linear model is  assumed for the signal, Moving Average (MA), AutoRegressive (AR), or AutoRegressive Moving Average (ARMA). Other common assumptions are linearity in general, as for the linearity of a system under inspection, and the Gaussianity of a signal, an assumption which greatly simplifies the calculations and leads often to linear equations. Moreover, the situation is that the fields of application of signal processing are so diverse and numerous that simulations or experiments performed to draw general conclusions as to when to use one estimation method or another are difficult to achieve, if not impossible. It is crucial for a reliable analysis to be able to quantify the effects of these assumptions and the choice of estimation method which both have on the final decision or judgement made by the observer. To do so, we begin by examining the main procedures followed traditionally in signal processing.

We start by recalling the merging point of the concepts of observation and information mentioned at the beginning of this chapter: That the observation must be coded in a certain **alphabet**. Then this observation will be a certain mathematical entity, belonging to a certain set or category. This automatically represents the observation in a certain context, i.e. a member in a set of broader extent. The other members of this set are considered to be "of the same kind" as the actual data, and these later being a particular "realisation". We recognise here the concepts of random variable and realisation of a random variable.

*Example 1: If the alphabet set is a number set, e.g. the real numbers, we have a real valued random variable.*

*Example 2: If we are dealing with logical circuits, their possible inputs or outputs, the random variable will have its values in the set of binary strings, i.e. sequences like 011011, eventually having a definite length or a maximum length.*

This view then leads us to one of the main tools of signal processing, which is probability theory. In this context, the observation , in its most simple form, is one trial of a random variable, the other members of the category being possible outcomes of the same random variable (r.v.).

The traditional practice followed in signal processing consists in looking at the first and second order statistics of the random variable or random process, that is, the mean and second moment for the case of a r.v.:

$$E(X) = m_1 \text{ and } E(X^2) = m_2$$

and mean and autocorrelation function for a random process:

$$E(X_k) = m_1(k), E(X_p X_{p+k}) = m_2(k)$$

for a stationary random process. However, according to the Kolmogorov theorem [Doob], [Gra], that a random process is specified in a unique way by the sequence of its multidimensional probability density functions

$$p(x_1, \cdots, x_n), n = 1, 2, \ldots$$

and, **in case these probability densities are well behaved functions**, namely, **analytic functions**, then they are determined by the infinite sequence of their moments

$$m_{i_1 i_2 \cdots i_n} = E(x_1^{i_1} x_2^{i_2} \cdots x_n^{i_n})$$

where $i_r$, $r = 1, \ldots, n$, are integers. Ignoring these higher order moments results in loss of information contained in the data. It could be argued that the information content of the moments is the same as the information content of the data, since, in general, the former are estimated from the latter. But this is not true. One point is that the moments are the coefficients of the Taylor expansion of the characteristic function which is positive definite [Rao], [BerCR] , a fact that imposes severe constraints on any would be sequence of moments. But it will be shown later in this chapter that the use of moments has deeper foundations than the positive-definiteness of the characteristic function, and this use of moments will still be important even when this latter constraint is relaxed or modified.

The action of taking the expectation value consists in contracting a multidimensional object (i.e. vector $v = [v_1, v_2, \cdots, v_p]^T$) or infinite dimensional object $f(x)$ to a scalar through integration: For the first case we use a scalar product

$$v.q = \Sigma v_i q_i \qquad\qquad (1.1.1)$$

and for the second case it is an integration

$$\langle f, q \rangle = \int f(x) q(x) dx \qquad\qquad (1.1.2)$$

Remarking that an expression like $x_1^{i_1} x_2^{i_2} \cdots x_n^{i_n}$ is in fact a concatenation of symbols, it can be concluded that taking the corresponding expectation value (and consequently the corresponding moment) is the act of drawing a conclusion from a set of observations, or, to be more precise, extracting the common features of these observed data. Now a question that may arise is about the case where the random variable is not numeric, a possibility that has been mentioned previously: For example, the case of random strings of characters, say *ab*, *ca*⟶*a*, *a+ccd*, etc., which may be found in a programming language, or more generally, in the grammar of a formal language. But here we can realise that this set of values can be mapped to a certain sufficiently large subset of the integers .

The picture now is the following: The information gained from observed data is necessarily represented by a p.d.f., this later being determined by the sequence of its moments - under the conditions previously stated, that they are analytic functions- which are features of observed data. This picture will allow us to formulate the problem in mathematical terms: Determining the p.d.f. $p(x)$ of a random variable $X$ from the sequence of its moments, and more precisely, from its moments up to a certain finite order $N$, $m_k, k = 1,2,\cdots, N$ and, further, estimation of higher dimensional p.d.fs of a random process from a sequence of their higher order moments. The study of this problem will be the subject of this thesis, and it will be shown, among other things, it will provide us with a unified view of the problems of signal processing and help in quantifying the effects of assumptions made and the choice of particular estimation methods on the outcome of the analyses, thus providing us with an efficient tool for reliable judgement and decision making.

For an observer (ultimately human observer, and may be an observer in the signal processing and control system sense) performing such a task, it (or he) gathers data, extracts the features (moments) and constructs a new object (p.d.f.) that is, an abstract construct defined by these and having a specific name, or, in other words, draws a "conclusion". The observer can then generate further examples of this new entity, or object, or decides whether a new set of observations belongs to the acquired category. These two activities are, respectively, extrapolation and recognition.

As examples of how a probability density can represent an observed object, we can cite the following:

*Example 3: For numeric data, the most elementary observation is a definite real number, arising from supposedly deterministic measurement (infinite precision): It can be represented as the $\delta$ function: Indeed, the density $p(x) = \delta(x-a)$ represents a deterministic measurement the outcome of which is the number $a$.*

*Example 4: If the deterministic condition in the previous example is relaxed, we will obtain an "approximate" value of $a$. In that case, a candidate distribution is a narrowband gaussian p.d.f. of mean $a$ and variance*

$$\varepsilon, 0 < \varepsilon << 1: p(x) = \frac{1}{\sqrt{2\pi\varepsilon}} \exp(-\frac{(x-a)^2}{2\varepsilon^2}).$$

*Example 5: As a more complicated example, consider the simple geometrical object, a straight line segment in the plane. The parameters describing it may be the co-ordinates of one extremity, the length $l$ and the direction, specified by an angle $\theta, 0 \le \theta < 2\pi$. We will allow, like the foregoing example, for a perturbation. Then what we will have is, say:*

$$p(x_1, y_1, l, \theta) = \delta(x_1 - x_{1_0}, y_1 - y_{1_0}) p_l(l) p_\theta(\theta)$$

*Where one extremity (the origin, of coordinates $(x_{10}, y_{10})$) was considered certain using the $\delta$ impulse function, and an uncertainty was allowed for the length and the direction through the use of the densities $p_l$ and $p_\theta$.*

An important point to note in the examples is that the difference between the use of the impulse deterministic distribution and the broader band one is not artificial, but rather reflects a deep and important characteristic of concept formation itself: A number, or any other object, is perceived by realising what it is and **what it is not,** that is, as a distinctive element in broader context, a fact which introduces the finite bandwidth densities rather than the impulse one, of zero bandwidth. For the last example, it was possible to introduce a probability density to reflect the origin, and even allow for deviations from the straightness condition, by introducing parameters of a polynomial, thus describing a curved line (around a mean, maximally likely, straightness, which is what is actually observed and supposed to reflect the reality).

## 1.2. Why the moments?

In the foregoing discussion, The moments were interpreted as representing common features extracted from the observed data. In this section, the role of the moments will be clarified and thoroughly justified, and shown to be inherent to the process of observation.

### 1.2.1. Maximum information retrieval

First of all, we recall that a probability density is determined in a unique manner by the infinite sequence of its moments in case it is analytic, a fact which is true for the solutions derived in this thesis as will be seen later. This is a mathematical fact arising from the Taylor expansion of the characteristic function

$$\Phi(\lambda) = \sum_{k=1}^{\infty} \frac{i^k m_k}{k!} \lambda^k \qquad (1.2.1)$$

and the one to one correspondence between a function and its Fourier transform

$$\Phi(\lambda) = \int p(x) \exp(i\lambda x) dx \qquad (1.2.2)$$

And therefore the information content of the p.d.f. lies in this whole sequence of moments. Ignoring the third and higher order moments causes eventual loss of information if the p.d.f. is estimated from data. On the other hand, the formulation of the problem in terms of **moments up to a certain finite order** will incorporate the limitations of the analysis carried out- such as the limited amount of data available- in the estimated p.d.f.. This later may be used to infer conclusions, performance indicators, and so on, and it is important that we are not going to derive unjustified inferences by "overloading" the data, or the p.d.f.. For the first point (loss of information), as an example, consider the two densities: The uniform distribution over an interval symmetric about zero

$$U(x) = \frac{1}{2a}, x \in [-a, a]$$
$$U(x) = 0 \quad \text{elsewhere}$$

and the zero mean Gaussian density

$$g(x) = \frac{1}{\sqrt{2\pi}\sigma} \exp(\frac{-x^2}{2\sigma^2})$$

The variance of the first density, the uniform, is

$$\frac{1}{2a} \int_{-a}^{+a} x^2 dx = \frac{1}{2a} \left[ \frac{x^3}{3} \right]_{-a}^{+a} = \frac{a^2}{3}$$

while, evidently, the variance of the Gaussian density is $\sigma^2$. If we chose the two parameters $a$ and $\sigma$ so that

$$a = \sigma\sqrt{3}$$

then the two random variables, corresponding to these two densities, have the same mean and the same variance, while they are certainly substantially different. For this particular example, we have even more: since the two densities are even, all the odd moments are zero. Clearly, we need the even order moments of order four and beyond to distinguish between them. The other point, that of inferring too much, will be studied and clarified in the following chapters, studying the various methods of estimation, and in a subsequent chapter comparing the results. Suffice to say here that, given the order $N$ considered, the moments of higher order will be functions of the first $N$ ones, so that we can say that they will be "tightly controlled".

### 1.2.2. Degree of detail

The analysis in the foregoing subsection relied on the main characteristic of the Taylor expansion of a function around some point, say, zero, as is the case here. This is the degree of approximation this limited expansion gives to the actual function. This degree of approximation follows the order of the limited expansion, or, in the terms of this thesis, the moments order considered. Since it has been proposed that an estimated p.d.f. represents a concept or an object, the higher the moments order, the more approximation we have to the object observed or "acquired". In other words, this order represents the degree of detail in the information about this object.

Now some calculus may clarify this assertion. If $p(x)$ is the estimated density, the moments are given by

$$m_k = \int x^k p(x) dx \quad k = 1, 2, \cdots$$

Choosing a "high" order for the estimation implicitly implies that the required p.d.f., $p(x)$, decays fast enough for large values of the variable $x$. That's because the larger the order $k$ is, the faster is the rate of growth and divergence of the monomial $x^k$, and we need the integral $\int x^k p(x) dx$ to be convergent (and equal to $m_k$). But the property of fast decay away from zero (or, equivalently, towards infinity) is nothing but the narrowness of bandwidth of the function considered: $p(x)$ is significant only in a relatively small interval. And all this is also true for bounded intervals and for expansion about an arbitrary point on the line, not necessarily zero. Therefore, if we chose a high moments order, we will get a narrowband p.d.f. In terms of objects, where this p.d.f. is supposed to represent a certain class, family or category, these latter will contain "few" elements: points where the p.d.f. is significant. And the converse is true: If we chose a low moments order, we'll get a wideband p.d.f., and the class represented will contain a substantial number of elements.

What do these observations have to do with the title of this subsection, the degree of details? It is that when an observer looks at minute details, differences will arise between objects that till now has been considered as alike, and so, in classifying observations, few objects will remain that have this large set of detailed common features. On the other hand, when considering, in the process of observations, a small number of details, and equivalently, ignoring further, more elaborate features, objects considered as different will, under this point of view, look similar. The final conclusion is that a high order moment estimation implies a high degree of details considered , and that a lower order is equivalent to ignoring finer details. This analysis will be recalled in a later subsection, where the concept of **complexity** is discussed.

For the case of numerical random variables, continuous or discrete (where the range of values is the set of integers), the foregoing analysis gives its own examples. Any function $f(x)$ decaying to zero no faster than $x^{-k}$ for $x \to \infty$ will not have moments of order $k-1$ or higher. In fact, for $x \to \infty$, $x^{k-1-\alpha} f(x)$ behaves like $x^{-1-\alpha}$, and we need the condition $\alpha > 0$ to insure the convergence of the integral.

An equivalent, but more insightful explanation, of the link between moment order and resolution or degree of detail is given by looking at the derivative of $x^k, k \geq 1$

$$d(x^k) = kx^{k-1}dx$$

so that for $x$ relatively large, $x > (1/k)^{\frac{1}{k-1}}$, we have that

$$\frac{d(x^k)}{dx} = kx^{k-1} > 1$$

that is, the higher is the order reached, the more substantial is the variation from one data item $x_i^k$ to another one, $x_j^k$. Whenever the observer considers that he has a "large" set of data, it is necessary to chose a higher order moment to resolve the differences between the elements of this set, which (the differences) must be considered as significant, or else the set wouldn't be large.

### 1.2.3. Shape and form

Having processed the observation following the moments approach proposed here, the observer will get a complete "picture" of the concept or object acquired. For a geometrical object, for example, like a circle, a square or a sphere, an observer will not perceive simply a set of points, a conclusion which would arise from an "only data" approach. Instead, he will observe a complete, single object (or, for more complicated data, a finite multitude of objects). The role of the moments in such a procedure lies in two characteristics:

It is a well established fact, in signal processing and even in modern physics, that the observables may be some kind of modal or **mean values** of more complicated or more detailed physical concepts, which, in some cases may be even considered as **unobservable**. This is the case for physical measuring devices, which always have a limited resolution in time or space, although this resolution may be high. And this should also be true for a human observer (for all senses, not only vision) since these senses perform their actions according to physical laws, this at least for a passive observation procedure. This point has been made at the beginning of this chapter when the duality observation-information has been noted, and this accounts for the signal processing case. But this is true also at a more fundamental level in quantum physics, where the **observables** are **expectation values** of some operators [Lin]. And we know that the moments are nothing but expectation values of elementary functions which are the monomials $x^p, p \geq 1$. These are elementary in the algebraic sense, since multiplication of two or more numbers is one of the elementary operations in algebra, and all actual calculations refer to it in case an evaluation is needed. But they are elementary also because of the next point.

The previous point stressed on the mean value characteristic of the moments. But it is possible to chose any set of functions, to find their expectation values. However, the problem now is there is a huge number of possible and acceptable functions, and what we were trying to avoid reappears: A set of different entities, unrelated in a clear way, which are the possible constraint functions. The particularity of the moments approach is that it is **recursive**: It calls itself. It is performed through the repeated (finitely many times) multiplication by the relevant variables and then taking the expectation values:

$$x \rightarrow E(x), xx = x^2 \rightarrow E(x^2), xxx = x^3 \rightarrow E(x^3)$$

and so on.

Now according to this view of the observation procedure, the set of points approach will be replaced be the mean or expectation value, a procedure which will be repeated recursively a finite number of times. And this is true for all the cases of observations (Although till now only numerical random variables have been considered). The outcome of all these discussions is that the moments approach proposed here will lead to a picture perceived one and in it whole completeness: The data, or points are smoothed through the act of taking mean values, and the geometric picture is constructed through the recursiveness. Then an observer can capture the picture, originally composed of many different items, as a whole one item. A complete expression of this point of view is made by the philosophical Aristotelian "matter" and "form", According to which matter can be conceived only when having a form, or shape[Ack]. At a more concrete level, some theories in cognitive science state that the human visual system perceives shapes and forms (the "Gestalt" school) [EysK], [Ver].

## 1.3. Complexity and duality

In this subsection, two important concepts will be introduced that will give deep insight into the problem of this thesis, that of estimating a p.d.f. from moments, and will complete the foregoing subsection in elucidating the special role of the moments. These two concepts are **complexity** and **duality**.

Complexity is a characteristic of a time series, or, more generally, of any finite family of observed data. It has more than one mathematical definition in the literature, and here we will use the version introduced in information theory by Kolmogorov [Kol1], which is the minimum length of a program which can generate this family of observations. The most simple example is a constant time series assuming the value of one, say:

$$x_k = 1, 0 \leq k \leq n$$

Then a code which can generate this sequence is

```
for k=0 to n
x(k)=1
```

21

```
end
```

if the number $n$ is relatively large, then this program is a more economic way to store or send the sequence, and we can say that this latter is of low complexity. More generally, if we have a periodic sequence of period $p$ and length $n$, then the sequence can be considered of low complexity as long as we have the condition

$$p \ll n$$

and in that case, a suitable program is

```
input a(0),a(1),...a(p-1)
k=0
while k+p<n+1
a(k+p)=a(k)
k=k+1
end
```

And finally, if we "perturb" a little the periodicity by altering a few terms of the series ("few" is taken comparatively to the number of periods), then we will have to add some code to specify each one of these values. The program will be a little longer, and accordingly the series is more complex.

On the other hand, duality will be introduced in mathematical terms, and is a characteristic of the relation between a numeric function, defined on the real line or on the circle, and its Fourier transform. It can be generalised to more abstract spaces through harmonic analysis. An overview of the main ideas and concepts in harmonic analysis and Fourier transform used in this thesis will be given later, but what is needed here is the "extent", or "bandwidth" of a numeric function. Given a function $f(x)$ on $\mathbb{R}$, and its Fourier transform $\hat{f}(\lambda)$, then if $f(x)$ is limited in extent, in the sense that it has significant value on a subset of $\mathbf{R}$ of small measure, then $\hat{f}(\lambda)$ has a large bandwidth, and vice versa. This is an important characteristic of the Fourier transform that is sometimes overlooked, and relates to the uncertainty principle in quantum mechanics. As an example of this fact, we can cite the impulse $\delta$ function, which is wholly concentrated at the point zero. Its transform is

$$\hat{\delta}(\lambda) = \int \delta(x)\exp(-i\lambda x)dx = 1$$

which is extended over the whole line. Another simple example is the square pulse of duration $\tau$

$$S(x) = 1, 0 \le x \le \tau$$
$$S(x) = 0 \text{ elsewhere}$$

Its transform is

$$\hat{S}(\lambda) = \frac{2\sin(\tau f/2)}{f}\exp(i\tau f/2)$$

a complex number, the modulus of which is

$$\frac{2\sin(\tau f/2)}{f}$$

For $f \approx 0$, that is, for $\dfrac{\tau f}{2} << 1 \Leftrightarrow f << \dfrac{2}{\tau}$, the value is $\tau$, a finite value. For substantially larger values of $f$, this value goes to zero. We conclude that the bandwidth of the transform is inversely proportional to the duration $\tau$ of the signal.

Consider now a set of observations $(x_k)_{1 \le k \le n}$ and a p.d.f. which may generate it, or that governs it, $p(x)$. If this p.d.f. is widely spread over the line, so that it takes significant values for a large set of values of the variable $x$, then it could be expected that a sequence generated by this p.d.f. will be very "complex", taking a large variety of values over its domain (through time or space, for example). This complexity is the same as the one introduced previously :A program that can generate the set $(x_k)_{1 \le k \le n}$ would be relatively long. On the other hand, if $p(x)$ is narrow band, then it has a definite pattern consisting of the values of the variable $x$

for which it is not negligible, a domain limited by the assumption of small bandwidth. A consequence of these remarks is that estimating a p.d.f. generating a set of data is in fact **a statement about the complexity of the sequence of data**. The role of the moments would be then to **express** or **implement** this statement. Recalling the examples given previously for the complexity concept, an ingredient will appear which is that the attributes "short" or "long" were used **with comparison** to the length of the sequence of data itself. Now given some sequence, suppose the observer judges it to be complex. This implicitly contains the fact that it is also a long sequence, of wide spread. Then although, being complex, the program or rule generating it would be long, it will be **valuable and short** compared to the spread of the sequence itself. In contrast, for a sequence with little complexity, a program generating it will be of limited value and long compared to the spread of the sequence. We recognise here the previously mentioned property of the Fourier transform, concerning the bandwidths of a transform pair. The important outcome of this discussion is the interpretation of the characteristic function of a p.d.f. as the rule or program generating the set of observations under consideration. In the language of computer science, the p.d.f., **estimated from its moments up to a certain order** provides the syntax, whereas the characteristic function is the **semantics.** It is important to remark that in this duality, the two members of the transform pair are equivalent. That is to say, only when we consider a function to be the original one it is eligible to say that the other one is the Fourier transform. Otherwise the latter would be the **inverse** Fourier transform and the first would be the **direct** transform. This is a characteristic of duality in mathematics and physics, and here it stems from the expressions of the direct and inverse transform pair:

$$\hat{f}(\lambda) = \int \exp(-i\lambda x) f(x) dx$$

and

$$f(x) = \frac{1}{2\pi} \int \exp(i\lambda x) \hat{f}(\lambda) d\lambda$$

These two expressions are identical, except for the minus sign in the first one, the interchange of which in the two expressions will give rise to the conjugation in the complex numbers set, an operation which is reflexive ($\overline{\overline{z}} = z$), a special characteristic of duality, and the factor $\frac{1}{2\pi}$, which is just a constant term, and has no effect formally. According to this view, the same will hold to the duality $syntax \longleftrightarrow$ $semantics$. And this conforms to reality, since the "semantics" of a text have to be coded in some specific form, an alphabet, which is again a syntax. Recalling again the duality $observation \longleftrightarrow information$ mentioned at the beginning of this chapter, it may be asked whether this latter has any thing to do with the duality we are talking about here. In fact, there's more than a simple occurrence of the concept of duality. What we consider as "observation" is the final step, the output and the apparent, whereas the information is the "inference", the conclusion and the internal. And it can be said that the first, the observation, is the syntax, and that the latter, the information, is the semantics.

The final remark in this section concerns the effects of applying the concepts of complexity and duality on "themselves". What can be said here is that complexity is clearly, by its very definition, an attribute of syntax only. It is true that it is defined by means of the program generating the data, but it is still the length of this program, obviously a syntax property, which is important in this definition. On the other hand, in the concept of duality is inherent the movement from one member of the duality pair to the other, so it doesn't stop at one side autonomously, in contrast to the complexity which keeps adherent to the syntax. In other words, **the complexity concept calls itself, and the duality concept is the act of calling.** This remark is the starting point for the process of analysing the methods proposed for estimation of a p.d.f., and it was felt necessary to discuss it in spite of its extreme abstractness.

## 1.4. Methods of estimation

The question now is how to proceed, given the moments in some way (from data, or through other ways, as will be shown later), to estimate the p.d.f..

Note first that the moments are derived from the p.d.f. by the relations

$$m_k = \int x^k p(x)dx, k = 1,2,\cdots$$

and that the c.f. is derived from the moments (**all** the moments) by the Taylor expansion

$$\Phi(\lambda) = \sum_{k=0}^{\infty} \frac{i^k m_k}{k!} \lambda^k$$

and we have that the p.d.f. and c.f. are a Fourier transform pair, so that **in principle,** the estimation of either of the two solves the problem. Therefore it can be expected, in the light of the previous section, that there may be two approaches: one building on the p.d.f., and thus corresponding to syntax and observation, the other one being performed by means of the c.f. and thus providing the semantics. The following analysis will show that this is true, and throughout this thesis it will be shown also that there's a complete duality between these two approaches.

At the beginning let's note that for either of the two approaches, the solution must be expressed in terms of optimisation of a certain criterion or principle. That's because we are considering a finite number of moments, and, although the choice of the higher order ones is restricted, it is not unique, so that we can expect to have many, and even infinitely many solutions. As usual in signal processing and control theory, the choice is performed through optimisation.

Now the first approach, supposed to deliver the solution p.d.f. directly, without resorting to the c.f., will start through the concept of complexity, as was discussed previously in this section and in the preceding section. We have a set of moments $(m_k)_{1 \le k \le N}$ representing some observed data, or obtained through another procedure and thus representing hypothetical data. The starting point is the assertion that the practical implication of building on the concept of complexity is that the complexity of the data sequence in hand is **typical**: When considered as a part of a larger sequence, this latter would have the same "pattern", or the same statistical characteristics (obviously up to a certain order, the moments order). The idea is that since the syntax was chosen to be the starting point, it should be assumed to contain everything needed. To say more, recall that the concept of complexity calls itself, and any additional syntax would be provided by calling external resources. Returning now to the choice of the principle governing the choice of solution, it must be phrased in terms of the given moments only (and certainly containing the required p.d.f), sticking to them without **calling** any other object. In other words, it must not involve any assumptions but those implied by the moments. This aspect of a principle is known to be the distinctive feature of **the maximum entropy principle**. The entropy of a random variable $X$ is given through its p.d.f. $p(x)$ by

$$H(X) = -\int p(x)\ln(p(x))dx$$

To maximise this function is equivalent to selecting the p.d.f. corresponding to the minimum of assumptions made , or containing **just the amount of information supplied by the moments, not more.**

We turn now to the other approach, supposed to proceed by means of the characteristic function. As was implied by the foregoing discussion concerning duality and complexity, this method will rely on the semantics, that is, on a certain understanding of the data represented by the moments. The operation will proceed as follows: The moments, constructing the characteristic function, are supposed to represent now the **semantics** of the observations. As it has been mentioned previously, these semantics in their turn will be expressed using certain **coding**, that is, some **syntax.** But the fact that the process is supposed to rely on duality, the semantics will be called again. In principle, this process will go on continuously, or to a certain number of times, in a manner that will be explained shortly. The important thing now is that when we go from coding to another, the later one **expressing the semantics** of the former, each word in the original code will be **explained**: It will be broken down into a more varied and numerous alphabet. The practical examples here are evident. A computer program written in a high level language will be compiled, and each instruction will be broken down into several ones to obtain a new code in assembler language, which in turn will be interpreted to machine level binary code. Beginning at a higher level, **macros,** consisting of a certain sequence of high level language instructions, but coded in one or only a few words, are another example. But these examples reveal an important feature of this process: Since the following code in every step uses a more varied and numerous alphabet set, there will be also different combinations of the letters and words of this alphabet giving other variants of the original **source code.** For each specific entity in this source code, a word, an instruction, or even a symbol, there will be other possibilities "around" it. For the same proposition, there are also examples for the case of numeric alphabets, i.e. the usual numeric random variables. If one observed value is $x$, supposed to be one dimensional for simplicity, it can be the outcome of a multidimensional system described by its state vector $z$ through the evolution equations

$$\frac{dz}{dt} = A.z + B.u$$
$$x = C.z \qquad\qquad (1.4.1)$$

where $A$ and $B$ are matrices with appropriate dimensions with respect to the state vector $z$ and the input, or control vector, $u$, and $C$ is a row vector. Since $z$ is multidimensional whereas $x$ is a scalar, the new "alphabet", represented by what now is considered to be the new observation, is more extended. This example is the most simple form of the Kalman filter.

The question now is what is the effect of this repeated sequence *coding* $\rightarrow$ *semantics* $\rightarrow$ *coding* on the data, moments, and the required solution. Talking in terms of the sequence of data, this repeated process will generate more and more **possible data** , considered as continuation of what is actually observed. As a more specific example, let's return to the symbolic random variables, citing a program composed of a sequence of instructions. One such instruction may be, say

```
c=a+b
```

If the + sign is interpreted as belonging to the category *symbol of a binary arithmetic operation*, we know that there are three other possibilities for this symbol: -, * and /; and if the category considered was enlarged to be *symbol of a binary operation*, then arise the alternatives of logical operators as well. Returning now to the main discussion, it is important to remark that these possible data are of higher complexity than the original set, and the continuation of this process will yield infinitely complex sequences, so that the process must be stopped at a certain stage. At this stopping point we have reached an agreed alphabet and syntax, but the effect on the p.d.f. (or ch.f.) estimation procedure was the following: At each step , the sequence obtained so far is looked at in the **context** of additional data, that is, **the statistics are modified to reflect the new point of view**. Now we proceed to the analytical description of this qualitative discussion.

Certainly, the point to begin with is the Taylor expansion of the characteristic function, the coefficients of which are given in term of the moments

$$\Phi(\lambda) = \sum_{k=0}^{\infty} \frac{i^k m_k}{k!} \lambda^k$$

if we have $\Phi(\lambda)$ we can recover the required p.d.f. using the inverse Fourier transform

$$p(x) = \int \exp(i\lambda x)\Phi(\lambda)d\lambda$$

But what we really have in hand are the moments only up to a certain finite order $N$, $(m_k)_{1 \le k \le N}$. We can think of considering the **truncated characteristic function**

$$\Phi_T(\lambda) = \sum_{k=0}^{N} \frac{i^k m_k}{k!} \lambda^k$$

and then finding the inverse transform. However, this is not feasible since the truncated c.f. is a **polynomial in** $\lambda$, **thus diverging at infinity**, so that it doesn't have a Fourier transform in the usual sense, the condition for the existence of this latter being the absolute integrability over the whole real line

$$\int_{-\infty}^{\infty} |u(x)| dx < \infty$$

Strictly speaking, it will have a transform in the sense of **distributions**, where this transform will be a combination of the impulse function $\delta$ and its derivatives. This aspect will be discussed at the end of this thesis, but it can be stated that in the context of this section, we should look for another way to obtain a solution. Since the problem is that of divergence at infinity, a possible way is to multiply the truncated c.f. $\Phi_T(\lambda)$ by another function decaying fast enough to zero for $\lambda \to \infty$. This function, $W(\lambda)$ will be called the weighting function.

Looking now to this analytical procedure in the light of the discussion preceding it, it can be realised that the divergence at infinity corresponds to the possibility of continuing indefinitely the operation

$coding \longrightarrow semantics \longrightarrow coding$

which will not give any useful result. **Now the weighting function is a smooth way for stopping this procedure**. All these points will be extensively discussed in chapter four. But for now, more aspects of the duality between the two approaches (syntax vs. semantics) will be elucidated that will give a more solid foundation and justification for their particular choice.

In an informal way, we can compare these considerations to a human observer perceiving the physical environment and acquiring information from this environment. One way to perform such a task is to acquire his information from previous, already learned ones. This step may certainly be repeated, and these previous ones could be considered as constructed from more elementary "building blocks", and so on. However, we should stop somewhere, at some stage of "resolution", and try to follow the way back to construct our really required solution. The analogy with the previous situation of coding and semantics is striking, and it may be anticipated that this decomposition into elementary constituents will lead again to the characteristic function approach. Indeed, at the final step, when it is decided that the most elementary components have been reached, any two of these ones must not have anything in "common" in terms of semantics, since otherwise the common object would be a "part" of these two, which violates the very fact that they are elementary. This conclusion is expressed in mathematical terms in the

**orthogonality of these elementary components**. We recognise here the decomposition of a function into its Fourier components, the complex exponentials $e^{i\lambda x}, \lambda \in \mathbf{R}$. This is certainly the characteristic function, the Fourier transform of the required p.d.f.. What has been done is that the unknown entity, the one to be acquired, has been put in a certain context, which is a set of orthogonal components or elementary building blocks. Again, we need to stop the decomposition at a certain stage, and this is the role of the weighting function. The required solution is now the projection of the weighted function on the space spanned by the elementary components. The projection, as known from linear algebra and analysis on vector spaces, involves minimisation of a distance, the proposed one for this particular method is the least squares distance, for reasons to be explained and discussed later. And this method will be called the context method, the least squares method or the projection method.

Alternatively, instead of building on previously learned and acquired objects, the required p.d.f. may be considered as a completely new entity, being itself, in a sense, a building block. This new entity can be learned by examples, which approximate it (the new concept or object). Since there will be no "construction" procedure, we expect that there will be no Fourier analysis, and the solution will be given directly in terms of a p.d.f., i.e. directly in **syntax** form. Moreover, being a new entity itself, the p.d.f. estimated by this method should have no correlation with any possible background, a fact which can be expressed precisely by saying that there will be no assumptions or external additional information involved in the estimation procedure. Again, this is evidently the characteristic property of the maximum entropy method.

Now the question, after all this analysis, is the actual **existence of a solution** in these two approaches. Certainly, this question will be addressed analytically and the solutions computed in the relevant chapters, but it is essential to clarify the very foundation of the existence of solutions, and to address this problem in a fundamental way. The strategy that will be followed will be founded on **the computability of the solutions** for both approaches, not the mere existence of such solutions (like, for example, the existence theorems of differential equations). That's because the problem dealt with is proposed to represent the link with the reality, so that the solution must be explicitly constructed and **computed**.

Recall the proposed justification and motivation for the problem of estimation of a p.d.f., which is to be able to quantify the effects of the assumptions made through the analysis of observations on the final judgement or decision taken by the observer. For a particular estimation problem and for a specific set of assumptions restricting the generality of this problem, the quantification may be achieved either through comparing theoretical results with real data, or by treating the problem more completely using more general and less restrictive conditions. However, in all this there are still subjective ingredients. These include the particular choice of what is supposed to be the real problem, and also the assessment of **how general is** this or the other particular approach. Now a procedure, a scheme or whatever is it, allowing us to treat **all** the possible situations must specifically exclude any possible object available to be interpreted as outside information or additional assumptions. The maximum entropy principle, formulated with the aid of moments, obviously has this characteristic. Including any additional assumptions is a deviation from this principle, and a measure of deviation would be valuable here. This introduces automatically the concept of distance, and suggests that the now modified principle which will be used to estimate the p.d.f. is based on a measure of distance. We can identify here what was called previously the minimum distance approach. In fact, while trying to find the solution, this later approach will still be tending to minimise the deviation from the reference point, the maximum entropy solution.

To begin the procedure, we'll start at the very beginning of any process: the coding. This point has been mentioned before, and here it will be looked at more closely. What happens in practice is that the observer chooses the particular alphabet from a relatively large set of possibilities, and chooses the "words" to be used, both choices being based on the subjective goal of the task and on somehow subjective conditions related to the expectations of the observer. But these two choices are nothing but relying on weighting the space of possible alphabets (using a probability density) and choosing the most likely alphabet; and the extent of detail, represented by the cardinality of the chosen set and the maximum word length corresponding to the maximum moment order (remarking that a $N^{th}$ moment order is a concatenation of $N$ symbols). It is possible to go back to a more fundamental level, and assert that the mere formulation of our problem as being the estimation of a probability density from its moments may be arrived at by looking at the common features of processes of learning and information extraction, and then building a general law (p.d.f.), we can infer three important points:

Our problem is recursive (It calls itself).

It is dynamic (Moving with the moments order).

The initial condition or starting point, which is necessary to choose properly in
    an adaptive process, calls the same problem.

All of these points will be looked at thoroughly at different places in this thesis, but here the important is the first point, recursivity.

This property, the recursivity as we have called it and as it is referred to in computer science and computation theory for a procedure which calls itself, is fundamental for the subject of this thesis. In computation theory, it corresponds to a function and routine which calls itself, and is common in some widely used high level languages (Pascal And C), and examples will be cited shortly. In philosophy, it is referred to as the "thing in itself", which we can explain, in our context, as the entity or object being considered as without giving any consideration to its actual value. In mathematics, it is represented by the formal variable in a polynomial or formal series. Finally, in technology, we propose it to be represented by an elementary frequency (or, may be, an element in an orthogonal set of components), and **acquired** by a phase locked loop. At a first glance, this type of construct may seem confusing or even paradoxical. In fact, one must be careful when invoking something of this kind. A rule that may be able to identify a paradoxical one is that it (the concept) **applies to itself**. An example is the Russell paradox [Penrose] , due to the mathematician and philosopher Bertrand Russell, which corresponds to "the set of all sets that aren't members of themselves". If this set is a member of itself, then it is not; and if it is not a member of itself, then it is. In contrast, a logical concept applies to real physical (in the broad sense) data and sets of data. It calls itself, but performing a repeated set of operations on **external** data. A typical example is the factorial function

$$n! = fact(n) = 1.2.\cdots.(n-1).n$$

A code implementing this function would be

```
function fact(n)
if n==0 then z=1
else
z=z*fact(n-1)
return(z)
```

Obviously, this function calls itself. However, the important thing is that the nested call applies to external data, the word "external" here meaning external to the structure of the procedure, in contrast to the previously mentioned, paradoxical concept which applies to itself. Another important feature is that these data can be enumerated so that the control is passed from the actual element to the next through a well defined procedure, that is, a recursive procedure. The final thing to note, from the example and the two foregoing points, is that the number of steps and calls in all this hierarchy must be finite **for every input element**, to ensure that the observer can have an outcome. The conclusion is that recursivity is equivalent to computability. These topics are well covered in the literature on computational theory, formal languages and finite state machines.

So now it is possible to call the problem of estimation of a p.d.f. from its moments is recursive, since it was shown to call itself in a sane way. But this is true, according to the previous discussion, for the maximum entropy method only. What can be said about the second method, the context or minimum distance method? In the brief discussion of this point, it has been mentioned that this procedure calls the concept of **deviation** and accordingly, the concept of distance. **Two** points will interact to clarify this type of recursivity. First, the fact the maximum entropy principle is a **reference point**, distinguished in the multitude of estimation procedures and inference strategies. The second point is that, through the act (discussed earlier in section $1.4$) of generation of a larger code space, i.e. splitting single letters in an alphabet to obtain many other, related ones, the concept of deviation will be invoked as an opposite alternative (in a would be two letters alphabet, or, equivalently, through the rule: character $\leftrightarrow$ opposite character) to the pointlike character of the maximum entropy principle. Calling this concept will destroy the uniqueness of maximum entropy principle and we'll obtain a new set of items, separated by distances, so that every item will be understood in the context of others. And the introduction of this context implies a scalar product as in equations (1.1.1) and (1.1.2)

$$\langle f | g \rangle = \int f(t) g(t) dt$$

Following which the distance introduced will be given by the quadratic form derived from this scalar product

$$(d(f,g))^2 = \|f - g\|^2 = \int (f(t) - g(t))^2 dt \qquad (1.4.2)$$

and, following this result, the context method, the characteristic function method, will be called the least squares method. To close this discussion of this method, it would be illuminating to go a little further in the analysis of the situation which arises when the deviation is introduced. We have seen that there will appear different entities, separated by distances, and understood in the context of each other. Applying this deviation to these ones will result in further fragmentation, and we must stop at a certain stage, where we will have elementary entities, and where it is not possible (practically, and in a certain specified setting) to obtain further reduction through scalar product. These are the elementary frequencies for the case of the real line $\mathbb{R}$, or the exponential functions $e^{i\lambda x}$.

The final point to be made before closing the chapter of duality consists in looking at it in the light of the theoretical and philosophical debate between two schools in probability theory: The first one lays the foundation of probability theory on the frequency of occurrence paradigm, which defines the probability in a certain **practical** situation according to the frequency of occurrence of each item in a sequence of observations (a fact equivalent to the law of large numbers). See [Mis]. The other one uses an axiomatic approach, relying heavily on advanced concepts in set theory and measure theory, and the law of large numbers is obtained as a **result** of this axiomatic approach. This school is sometimes called the "subjective" view. In our context here, the first view should correspond to the syntax and complexity based maximum entropy method, whereas the second view (the subjective one) corresponds to the semantics, context method.

## 1.5. Origin of the moments

It has now become clear that the role of the moments is a cornerstone in the estimation problem in this thesis, this later being formulated in terms of these moments; and this formulation has gained extensive analysis and justification in the preceding sections. The question now is that, knowing that the moments are the starting point of the estimation procedure, how can we have access to them? One evident and well known way is to get the moments from the observed data. That is, given a series of observations $(x_k)_{1 \le k \le n}$, the moments are estimated from these data through the relations

$$m_p = \frac{1}{n} \sum_{k=1}^{n} x_k^p, p = 1,2,\cdots \qquad (1.5.1)$$

However, this doesn't have to be always the case. In fact, it would be very limiting and restrictive for the theory presented in this thesis that the only source of moments is through mean values of data. The idea is based on the fact that a p.d.f. describes the whole properties of the system. In cases where we believe that we have a whole description of a system, we may have access directly to its whole p.d.f.. Such a situation may arise whenever we have a **state space** description of a system, which evolves through time governed by what is known as a **stochastic differential equation**. Recall the system described by equations (1.4.1), the Kalman filter, but with a slight change

$$\frac{dx}{dt} = ax + bu \qquad (1.5.2)$$

where the observation equation has been omitted, and the scalar $x$ (for simplicity) is the state of the system. Assume that the driving input $u(t)$ is a random process. Then the state $x$ will be also a random process. If we take the expectation values at both sides we obtain

$$\frac{dm_x}{dt} = am_x(t) + bm_u(t) \qquad (1.5.3)$$

an equation governing the evolution in time of the first moment of the state variable $x$. In practice, the driving input random noise would be a process of relatively simpler structure and properties. In fact, it may be the goal of the very model to infer such basic driving "background". This is the case, for example, in spectral factorisation, where the aim is to find a model for a stationary random process as the output of a linear system driven by a white noise.

Now the higher order moments are in terms of the powers of $x$, and we have that

$$\frac{d(x^k)}{dt} = kx^{k-1}\frac{dx}{dt}$$

so that, taking expectations at both sides

$$\frac{d(m_x)_k}{dt} = E\left\{kx^{k-1}\frac{dx}{dt}\right\} = E\left\{kx^{k-1}(ax+bu)\right\} = E\left\{akx^k\right\} + E\left\{bkx^{k-1}u\right\}$$

The final result is that

$$\frac{dm_{xk}}{dt} = kam_{xk} + bkE\left\{x^{k-1}u\right\} \qquad (1.5.4)$$

In the case where $x$ and $u$ are uncorrelated, we obtain

$$\frac{dm_{xk}}{dt} = kam_{xk} + kbm_u m_{(k-1)}$$
(1.5.5)

As long as the approximations and assumptions are valid, we can construct the whole p.d.f of the state variable $x(t)$ at subsequent times, away from the starting point.

A similar example, but from another point of view, is to consider the same system (1.5.2), but with no random driving

$$\frac{dx}{dt} = ax$$
(1.5.6)

Instead, the uncertainty is in the **initial condition**, $x(0) = x_0$, now considered as a random variable obeying a certain distribution given by the density $p_0(x_0)$, which may be available either explicitly, or through a number of its moments. Certainly, the second case is feasible relying on results of this thesis. Either way, repeating the operations done on the previous example, we have, for each moment of the state variable $x(t)$

$$\frac{dm_k}{dt} = kam_k$$
(1.5.7)

The two systems considered are particular cases in a more general setting, given by the Markov processes theory. Here, the transition probability density of a Markov process is governed by non-linear partial differential equation, known in two variants of it as the Fokker-Planck equation and the Kolmogorov equation [Dyn]. This equation involves the first order time derivative of transition p.d.f, and higher order partial derivatives of this p.d.f with respect to the vector state variable, as well as functions which are in a sense, the higher order moments of the increment process $\Delta x(x,t)$. Combinations of experimental moments estimation, rephrasing of the p.d.f in terms of its higher order moments, differential equations techniques and finally estimation of a p.d.f. along the line proposed in this thesis will yield powerful techniques for studying and simulations of such systems, known to have great relevance to a large multitude of applications.

## 1.6. Background

The background to the material of this work is not extensive, but if we relax the restrictions specifying this work precisely (Estimation of a p.d.f. from its moments up to a certain order), it is possible to find varied aspects discussed in a certain number of previous publications, including both papers and monographs.

In fact, we can look at the theme of this thesis as belonging to the now mature field of higher order statistics; and we can look at the term "estimation of a probability density", where we can find diverse works on this topic, especially research papers. Finally, the "moment problem" has for long been studied from the pure mathematical viewpoint by famous mathematicians, and a wealth of important theoretical results has been reported. On the other hand, in contrast to this historical background, we can speak of "logical" background, that is, of methods which are best viewed as "precursors" to the approach adopted in this thesis; these are parameter estimation and the empirical distribution function. We'll begin by this logical background.

### 1.6.1. Logical background

One usual topic in statistics is of determining a probability density which depends on a certain number of parameters. The estimation is based on experimental observed data, and this problem is classified as pertaining to hypothesis testing. The treatment of this problem is abundantly found in statistics literature, where we can cite the modern text [StuO]. In [Sor], the general problem of parameter estimation is studied, but it readily becomes specified in terms of what is meant here, of estimation of the parameters of a distribution or a density. It will be shown later, in chapter four, that this topic can be classified as a particular setting in the general method of least squares. It is important to mention that a special method of parameter estimation which is relevant to the main theme of this thesis, and it is known as the moments method. In this method, the moments are estimated experimentally from data, and then from the model distribution, and a system of equations is solved for the parameters. This method is analysed fairly thoroughly in [StuO], as well as in [Sor], where it is reported to be of poor performance.

The other precursor of the theme of p.d.f. estimation is known in literature as the empirical distribution function (edf). In this method, a distribution is constructed directly from the (empirical) data; its principle is that every item of data is assigned a probability proportional to the frequency of its occurrence in the sequence of data. This foundation for the method is intuitive, and it is not clear when it first did appear. However, the convergence theorem (given in chapter 3) is attributed to Glivenko, or Glivenko and Cantelli. A method for density estimation, due to Parzen, appears in the paper [Par]. In the paper [Mas1], a method for p.d.f. estimation based on the method of Parzen is used. Finally, we can cite the monographs [Ahi] and [SohT] as pertaining to the empirical distribution approach, although these works address the mathematical moment problem. The reason for their inclusion is that the edf is proposed as a solution to the moment problem. In this connection, a newer, and mathematically advanced treatment of the moment problem is found in [BerCR].

## 1.6.2. Historical background

As a historical background, we have two categories of research work. One, and may be the most important and more relevant to the subject of this thesis, is the work on the now nearly mature field of higher order statistics, and this thesis can be considered as a contribution to this field, but with some particular and decisive and important differences. The other category is the previous work on "estimation of a probability density".

Among research work on estimation of a p.d.f., there are a few papers addressing this problem under exactly this heading, but none of them mentions the moments as a special, or even possible tool. In the paper [LeiR] a Bernoulli distribution (Two possible values of the density: $p$ and $q = 1 - p$) is estimated using a least squares error measure and finite memory, that is, a finite number of samples. The paper [Par] is used in this thesis as a base for a density estimation method corresponding to the edf. We cite two papers for Masry, where [Mas1] has been mentioned previously as building on the method of Parzen, and [Mas2] which considers the problem of estimation of the joint multidimensional densities of a stationary random process, from noisy observations. It builds on a mixture of [Par] and deconvolution methods. An important approach is carried out in [BarC], where the estimation of a density is based on the Kolmogorov complexity, and the solution is to selected from a prescribed set of possible p.d.fs. In this paper, the connection between complexity and the estimation of a p.d.f. is remarked and discussed, closely to the important conclusion of this thesis on this connection.

The higher order statistics (HOS) aspect of the background may be more relevant, at least in the basic idea of considering higher order moments, as is the approach in this thesis, or higher order cumulants, which is the adopted approach in the vast majority of research works in this field. A relatively new textbook on the conventional treatment of HOS is [NikP]. The paper [Men] offers a complete view of this field, including possible applications. As for specialised applications, we can cite typical papers addressing two particularl applications of HOS: Identification of nonminimum phase systems and blind deconvolution. The former one is based on a particularity of higher order spectra, which is that they are phase sensitive, in contrast to the second order power spectrum; here we can mention the paper[GiaM]. The latter is based on the ability of higher order spectra to retrieve much more information from data, and two example papers are [Petr] and [Jar].

# CHAPTER 2

# PRELIMINARIES

In this chapter, some basic mathematical results from various domains in mathematics are gathered, which may be needed later in the bulk of this thesis. The treatment will cover main concepts in probability theory, beginning with random variables, then follows a discussion of random processes. Topics in functional analysis, namely function spaces and optimisation, are addressed. Finally, main concepts in harmonic analysis, of great importance in the thesis, are reviewed. Although most of the material in this chapter is an account of already established main results in the mentioned topics, at some places the discussions are oriented towards the subject of the thesis, and some new points of view may arise.

## 2.1. Random variables

We begin in this section by reviewing the main foundations of the axiomatic theory of probability.

Given a set $\Omega$, a $\sigma$-algebra $\Sigma$ of sets in $\Omega$ is a collection of subsets of $\Omega$ satisfying the following properties:

1. If the set $A \in \Sigma$, then also $A^C \in \Sigma$, where $A^C$ is the complement of $A$ in $\Omega$.

2. If we have a sequence of subsets of $\Omega$, $(A_i)$, $i = 1,2,\cdots$, (not necessarily a finite sequence), then $\bigcup A_i \in \Sigma$.

3. $\emptyset \in \Sigma$

These properties imply that $\Sigma$ is closed under the operations of union, intersection, and difference of sets.

A measure on the space $(\Omega, \Sigma)$ is a mapping

$$\mu{:}\Sigma \to \mathbf{R}^{+}$$

which is countably additive, i.e. for $(A_i)$, $i = 1,2,\cdots$, a sequence of mutually disjoint subsets in $\Omega$, we have that

$$\mu(\bigcup A_i) = \Sigma \mu(A_i) \tag{2.1.1}$$

$\mu$ is called a probability measure if

$$\mu(\Omega) = 1$$

A space equipped only with a $\sigma$-algebra (without specifying any particular measure) will be called a measurable space.

Sets in a family of sets in $\Sigma$, $(A_i)_{1\leq i \leq N}$, are said to be mutually independent if for each sequence $(i_k)_{1\leq k \leq m}, m \leq N$, we have

$$P(\bigcap_{k=1}^{m} A_{i_k}) = \prod_{k=1}^{m} P(A_{i_k})$$

Given a probability space $(\Omega_1, \Sigma_1, P_1)$ and a measurable space $(\Omega_2, \Sigma_2)$, a map $X{:}\Omega_1 \to \Omega_2$ is said to be measurable if, for every $S \in \Sigma_2$, $X^{-1}(S) \in \Sigma_1$. In that case, $X$ is called a random variable. In this thesis, we consider mainly $\mathbf{R}^d$-valued random variables, where $\mathbf{R}^d$ is equipped with the $\sigma$-algebra of its Borel sets, which is the smallest $\sigma$-algebra containing the open sets. Having that, the mapping (random variable) $X$ will induce a probability measure $P_2$ on $\Omega_2$ defined by: $P_2(A) = P_1(X^{-1}(A))$. Then $P_2$ is called the distribution of $X$. In the particular case where $\Omega_2 = \mathbf{R}^d$, the name "distribution" will be reserved for the function

$$F(x_1,\ldots,x_d) = P_1\{X_1 \leq x_1,\ldots,X_d \leq x_d\}$$

If there's a function $p(x_1,\ldots,x_d)$ such that

$$F(x_1,\ldots,x_d) = \int_{-\infty}^{x_1} \cdots \int_{-\infty}^{x_d} p(t_1,\ldots,t_d)dt_1\cdots dt_d$$

that is, a density for this distribution, then this density will be called the probability density of the corresponding r.v.. In the one dimensional case we have

$$p(x) = \frac{dF(x)}{dx}$$

Then for real valued r.vs. we define the mean, or expectation of the r.v. $X$ to be

$$E(X) = \int xp(x)dx = m_1$$

the first moment. The $k^{th}$ order moment of $X$ is

$$m_k = E(X^k) = \int x^k p(x)dx$$

We define next the characteristic function (c.f.)of $X$

$$\Phi(\lambda) = \int \exp(i\lambda x)p(x)dx$$

It is the Fourier transform of the p.d.f. Then we have the Taylor expansion of this ch.f.

$$\Phi(\lambda) = \sum_{k=0}^{\infty} \frac{i^k m_k}{k!}\lambda^k$$

These results are easily generalisable to the multidimensional case:

$$\Phi(\lambda_1,\ldots,\lambda_n) = \sum_{k_1,\ldots,k_n=0}^{\infty} \frac{i^{|k|} m_{k_1\cdots k_n}}{k_1!\cdots k_n!}\lambda_1^{k_1}\cdots \lambda_n^{k_n}$$

where

$$|k| = k_1 + k_2 + \cdots + k_n$$

and

$$m_{k_1\cdots k_n} = E(X_1^{k_1}\cdots X_n^{k_n})$$

for a $\mathbf{R}^n$ valued random variable.

*Remark: The Fourier transform used to define the characteristic function in probability theory is the inverse fourier transform in other fields of applications:*

*Note that there's no minus sign in the exponential in the definition of the characteristic function. However, we will use the nomination "Fourier transform" rather than "inverse Fourier transform" according to the current practice in almost all literature on probability theory.*

We adopt the usual known concept of independence without unnecessary mathematical formulation. Then for mutually independent r.vs. $X_1, \ldots, X_n$, the joint multidimensional p.d.f. is the product of the individual p.d.fs.

$$p(x_1, \ldots, x_n) = \prod_{i=1}^{n} p_i(x_i)$$

The second ch.f. is defined as

$$\Psi(\lambda) = \ln \Phi(\lambda)$$

Now if $X_1$ and $X_2$ are two independent random variables, then the distribution of their r sum is given by

$$p_{X_1 + X_2} = p_{X_1} * p_{X_2}$$

that is , it is the convolution of the two individual densities. This leads to the relations

$$\Phi_{X_1 + X_2} = \Phi_{X_1} \Phi_{X_2}$$

and

$$\Psi_{X_1 + X_2} = \Psi_{X_1} + \Psi_{X_2}$$

Expanding $\Psi$ into its Taylor series

$$\Psi(\lambda) = \sum_{k=1}^{\infty} \frac{i^k \mu_k}{k!} \lambda^k$$

then the coefficients $\mu_k$ are called the cumulants of the corresponding r.v.. A general relation between cumulants and moments can be worked out from a formula in [NikP] considering the general multidimensional case of a random process. We have

$$\mu_n = \sum_{p=1}^{n} (-1)^{p-1}(p-1)! \sum_{k_1+\cdots+k_p=n} m_{k_1}\cdots m_{k_p}$$

Consider now a sequence $(X_n)_{n=1,2,\ldots}$ of r.vs. defined on the probability space $(\Omega, \Sigma, P)$. There are many ways in which convergence of this sequence to a r.v. $X$ can be defined [Doob], [Rao]. Here the strongest type of convergence, convergence almost everywhere (a.e.): $X_n \to X$ a.e. if there's a set $N \in \Sigma$, $P(N) = 0$, and $X_n(\omega) \to X(\omega)$ for $\omega \in \Omega - N$. Having that, we state now a form of the law of large numbers which will be the base for estimating moments from data: If $(X_n)_{n=1,2,\ldots}$ is a sequence of independent identically distributed random variables, of mean $m$ and variance $\sigma < \infty$, then

$$\frac{1}{n}\sum_{k=1}^{n} X_k \to m \text{ a.e. as } n \to \infty$$

Now given a sequence of trials of a single r.v. $X$. it can be considered as a sequence of realisations of i.i.d. r.vs. $x_1,\ldots,x_n$, and according to the previous result, the law of large numbers, an estimate of $m_k = E(X^k)$ is

$$\hat{m}_k = \frac{1}{n}\sum_{i=1}^{n} x_i^k$$

for large $n$. Certainly, the accuracy of this estimation depends on how large is $n$, and how large is it compared to the variance $\sigma$. We can calculate the variance of this estimate by considering it as a random variable

$$S_{n,k} = \frac{X_1^k + \cdots + X_n^k}{n}$$

We have that $E(S_{n,k}) = \dfrac{1}{n}\sum_{i=1}^{n} X_i^k = \dfrac{nm_k}{n} = m_k$, since the random variables $X_i$ are identically distributed. Let now $\sigma_{n,k}^2 = E(S_{n,k} - m_k)^2$ be the required variance. Developing the expression, and noting the independence of the $X_i$, we obtain

$$\sigma_{n,k}^2 = \frac{m_{2k} - m_k^2}{n}$$

This quantity depends on the p.d.f., and there are cases where it is not sufficiently small. Suppose, for example, that the p.d.f is symmetric about zero, that is, even. Then $m_k = 0$ for $k$ odd, and $m_{2k} = \int_{-\infty}^{\infty} x^{2k} p(x)dx > 2\int_{\beta}^{\infty} x^{2k} p(x)dx > \beta^{2k} A$, where

$A = 2\int_{\beta}^{\infty} p(x)dx$ and $\beta > 1$. If $A > 0$ then $m_{2k} \to \infty$ as $k \to \infty$, and so does $\sigma_{n,k}$

since, for $m_k = 0$, $\sigma_{n,k}^2 = \frac{m_{2k}}{n}$. This behaviour persists, by continuity, if the p.d.f. is a little asymetric about zero. It can be seen that in such cases, we must consider the moments only up to a certain order where this variance is still tolerable.

A probability density which is of interest is the Gaussian or normal density, and is given by

$$p(x) = \frac{1}{\sqrt{2\pi\sigma^2}} \exp(-\frac{(x-m)^2}{2\sigma^2})$$

It is of mean $m$ and variance $\sigma$. Its c.f. is

$$\Phi(\lambda) = \exp(im\lambda - \frac{\sigma^2 \lambda^2}{2})$$

If $\Phi(\lambda)$ is expanded in its Taylor series, which amounts to the expansion of an exponential function, we can see that the coefficients of $\lambda^k$, $k \geq 3$ are polynomials in $m$ and $\sigma$, so that the higher order moments are functions of the first two. If we look now at the second ch.f.

$$\Psi(\lambda) = im\lambda - \frac{\sigma^2 \lambda^2}{2}$$

we see that all the cumulants of order three or more are zero.

## 2.2. Random processes

A random process is a family $(X_t)_{t \in T}$ of r.vs over the space $(\Omega, \Sigma, P)$, where $T$ is the observation time window, and could be, in the continuous time case, an interval or the whole real line, and in the discrete parameter case, a sequence of integers. For each $\omega \in \Omega$, the function of $t$, $X_t(\omega)$, is called a realisation or sample function of the random process.

For each finite sequence $t_1, \ldots, t_n \in T$, consider the $n$-dimensional distributions of $X_{t_1}, \ldots, X_{t_n}$ given by

$$F_{t_1, \ldots, t_n}(x_1, \ldots, x_n) = P\left[ X_{t_1} < x_1, \ldots, X_{t_n} < x_n \right]$$

They satisfy the following consistency conditions [Doob]:

$$\lim_{x_n \to \infty} F_{t_1, \ldots, t_n}(x_1, \ldots, x_n) = F_{t_1, \ldots, t_{n-1}}(x_1, \ldots, x_{n-1})$$

$$\lim_{x_n \to -\infty} F_{t_1, \ldots, t_n}(x_1, \ldots, x_n) = 0$$

and if $(i_1, \ldots, i_n)$ is a permutation of $(1, \ldots, n)$

$$F_{t_{i_1}, \ldots, t_{i_n}}(x_{i_1}, \ldots, x_{i_n}) = F_{t_1, \ldots, t_n}(x_1, \ldots, x_n)$$

The corresponding p.d.fs satisfy the corresponding conditions deduced by integration over the variables. The Kolmogorov extension theorem [Doob],[Gra], states that the converse is true under mild conditions which are almost always met in practice, that is, given functions satisfying the above conditions, there is a random process having these functions as its multidimensional distributions. Moreover, from these distributions we can construct a probability distribution on $\mathbf{R}^T$, the space of real valued functions on the interval $\mathbf{T}$, or on the whole line. In this setting, $\mathbf{R}^T$ is equipped with a suitable $\sigma$-algebra, the product $\sigma$-algebra. Thus a random process is uniquely determined by the family of its finite dimensional distributions.

## 2.3. Function spaces

Let $S$ be a vector space. A norm on $S$ is a map $N:S \to \mathbf{R}^+$ satisfying the following properties:

$N(x) = 0$ if and only if $x = 0$.

$N(x+y) \le N(x) + N(y), x,y \in S$.

$N(\alpha x) = |\alpha| N(x), \alpha \in \mathbf{R}$.

The norm of a vector $x$ is denoted $\|x\|$. A scalar product on $S$ is a positive definite symmetric bilinear form $B$:

1. $B(x,y)$ is linear in $x$ and $y$.

2. $B(x,y) = B(y,x)$.

3. $B(x,x) > 0$ if $x \ne 0$.

Then $\|x\| = \sqrt{B(x,x)}$ is a norm on $S$. We'll use the notation $B(x,y) = \langle x,y \rangle$. The space $S$, equipped with this norm, is called a pre-Hilbert space. If, in addition, $S$ is a Banach space, then it is called a Hilbert space. Two elements in $S$ are said to be orthogonal, and we write $x \perp y$, if $\langle x,y \rangle = 0$. Let $M$ be a subset of $S$. We denote by $M^\perp$ the subset of $S$ containing the elements orthogonal to every vector in $M$. Then we have the following result [Yos]

*Theorem*: Let $M$ be a closed linear subspace of the Hilbert space $S$. Then $M^\perp$ is also a closed linear subspace, called the orthogonal complement of $M$, and any vector $x$ in $S$ can be decomposed in a unique way in the form $x = m+n$, where $m \in M$ and $n \in M^\perp$. Moreover, we have

$$\|n\| = \|x - m\| = \inf_{a \in M} \|x - a\|$$

so that $m$ is the closest vector to $x$ in $M$.

This theorem is the basis for the least squares method, of common use in signal processing. For a more general Banach space, we don't have a similar result, because we don't have the concept of orthogonality. However, the essence of the foregoing theorem is in minimising the distance function, and, in fact, we can perform optimisation of functionals on general Banach spaces. The difference is that the solution will not be necessarily unique [Yos], and in that the available norm may not be amenable to the operation of differentiation, an operation that can be performed on quadratic forms.

The function space to be used in this thesis will be the space of real or complex valued, square integrable functions on the real line, usually denoted $L^2$. The usual norm on this space is given by the bilinear form

$$\langle f, g \rangle = \int_{-\infty}^{\infty} f(t)\overline{g(t)} dt$$

accordingly, the norm will be given by

$$\|f\| = (\int_{-\infty}^{\infty} |f(t)|^2 dt)^{\frac{1}{2}}.$$

## 2.4. Optimisation

Consider a Banach space $\mathbf{S}$, and a continuous map $F : \mathbf{S} \to \mathbf{R}$. $F$ is said to be Frechet derivable [IofT] at $x \in \mathbf{S}$ if there is a linear functional $L_x$ on $\mathbf{S}$ such that

$$F(x+h) - F(x) = L_x.h + o(\|h\|)$$

This is an extension to the differentiation on $\mathbf{R}^n$. Now a fundamental result is that [IofT] a necessary condition for $F$, supposed to be Frechet derivable on $\mathbf{S}$, to reach an extremum at $x \in \mathbf{S}$ is that we have $L_x = 0$. As an example, let's find the derivative of the square of the norm function $\|x\|^2$ in a Hilbert space. We have

$$\|x+h\|^2 - \|x\|^2 = \langle x+h, x+h \rangle - \langle x, x \rangle = \langle x, h \rangle + \langle h, x \rangle + \langle h, h \rangle =$$
$$\langle x, h \rangle + \langle h, x \rangle + \|h\|^2 = \langle x, h \rangle + \langle h, x \rangle + o(\|h\|)$$

so that the derivative is the linear functional

$$L_x . h = \langle x, h \rangle + \langle h, x \rangle$$

In the case of a real valued scalar product, this is $2\langle x, h \rangle$. For a complex valued scalar product (Hermitian form), it is $2\,\mathrm{Re}\langle x, h \rangle$, Re being the real part of the complex number.

In the majority of cases, we have constraints: The vector $x$, at which the functional $F$ reaches its extremum, is subject to certain constraints of the form

$$f_i(x) = 0, \quad i = 1, \ldots, n$$

In that case, we have the well known result, stated for the sake of completeness, that at the solution point to the problem with constraints, $x^*$, the Lagrange function

$$F_\lambda(x) = F(x) + \sum_{i=1}^{n} \lambda_i f_i(x)$$

must have a null Frechet derivative $F_\lambda'(x^*) = 0$.

## 2.5. Harmonic analysis

The discussion in chapter 1 has shown great emphasis on the Fourier transform and harmonic analysis in general, in that the latter topic studies the decomposition of mathematical or physical entities (represented mathematically) into their basic ingredients and the reconstruction of these entities. The basics of harmonic analysis will be presented in this section, stressing those concepts that have been (and will be) mostly used in this thesis. For a more complete and elaborate treatment, see [HewR].

The Fourier transform of an absolutely integrable function on the line, i.e. satisfying

$$\int_{-\infty}^{\infty} |f(t)| dt < \infty$$

is given by

$$\hat{f}(\lambda) = \int_{-\infty}^{\infty} \exp(-i\lambda t) f(t) dt$$

On the Hilbert space of square integrable function on the line, the inverse transform always exist and is given by

$$f(t) = \frac{1}{2\pi} \int_{-\infty}^{\infty} \exp(i\lambda t) \hat{f}(\lambda) d\lambda$$

and this mapping preserves the inner product and is an isometry

$$\int_{-\infty}^{\infty} |f(t)|^2 dt = \int_{-\infty}^{\infty} |\hat{f}(\lambda)|^2 d\lambda$$

Mathematically, the Fourier transform can be defined on a general algebraic structure which is a topological group. It is well developed especially on locally compact, commutative groups and compact, not necessarily commutative groups. An example of the former is the one just considered, the real line $\mathbf{R}$. An example of the latter, for the commutative case, is the circle $\mathbf{T}$, which corresponds to the well known Fourier decomposition of periodic functions. In all these cases, the domain of the Fourier transform is a dual group, which, for the mentioned two examples, is again the real line $\mathbf{R}$ for the former, whereas it is the integers set $\mathbf{Z}$ for the latter. In the general case of a commutative locally compact group, the dual group is the "character group", which is the group of multiplicative homomorphisms from the original group into the field of complex numbers $\mathbf{C}$. One main property of the Fourier transform, which is its main reason to be used in applications, is that it transforms convolution, which is a fairly complicated mathematical operation, to mere multiplication. If we have

$$h(t) = (f * g)(t) = \int_{-\infty}^{t} f(u)g(t-u)du$$

then , in the Fourier domain

$$\hat{h}(\lambda) = \hat{f}(\lambda)\hat{g}(\lambda)$$

Another property, which has been of importance in the discussions in the first chapter, is the "uncertainty" relation, which states that the product of the bandwidths of a function and its transform has a minimum, so that they can't have both in the same time a large bandwidth or a small bandwidth. We have [DymM]

$$\int x^2 |f(x)|^2 dx \times \int \lambda^2 |\hat{f}(\lambda)|^2 d\lambda \geq A\|f\|^4$$

Used with the following property of the Fourier transform, this result can be extended to higher powers in the variable, $x$ or $\lambda$. The property is that, for a Fourier transform pair $f(x)$ and $\hat{f}(\lambda)$

$$F\left\{\frac{d^p f(x)}{dx^p}\right\} = i^p \lambda^p F\{f(x)\} = i^p \lambda^p \hat{f}(\lambda)$$

$$F^{-1}\left\{\frac{d^p \hat{f}(\lambda)}{d\lambda^p}\right\} = (-1)^p i^p x^p f(x)$$

Suppose now that the function $f(x)$ is very "flat" , so that its derivatives up to order $p$ are almost constant on a certain interval, sufficiently large (the more large is $p$, the more large is the interval). Then the bandwidth of $\dfrac{d^p f}{dx^p}$ is large, and by the foregoing result on the bandwidths of a transform pair, its Fourier transform, $i^p \lambda^p \hat{f}(\lambda)$, is of restricted bandwidth. But here we have the function $\hat{f}(\lambda)$ multiplied by the power $\lambda^p$, which grows fast for large values of the variable. That means that $\hat{f}(\lambda)$ decays very fast at infinity, in fact as fast as the function $f(x)$ is flat, or steady. The converse type of behaviour is also true because of the second one of the equations above stated. This shows the complete duality in behaviour and clarifies this relationship between the bandwidths of a Fourier transform pair, a fact used extensively in the discussions in the first chapter.

Fourier analysis can be interpreted as the decomposition of an object into its main constituents, which are the individual frequencies. If this interpretation is to be true, there should be a question of synthesis, that is, reconstructing an object from these constituents. In fact, the problem of spectral synthesis is a classical problem in the field of harmonic analysis, and will be of special interest in this thesis. We will here outline briefly the principal ideas and results relevant to the discussions in the thesis. The abstract language of groups will be used and the statements will be rather heuristic. For a complete mathematical treatment, see [HewR].

The spectrum of a function $f(x)$ in $L^1(G)$ is the set of points in the dual group $\hat{G}$ for which the Fourier transform $\hat{f}(\lambda) \neq 0$. Given now a set $S$ in $\hat{G}$, we can look at the functions in $L^1(G)$ whose spectrum lies inside $S$. The question is: Is it possible to construct these functions from their spectrum? The first point is that there are many ways of summation, or weighted summation, of individual components in $S$. A "way" is described by a weighting function which assigns a weight to each item in the sum, or more generally, by a measure on the set $S$:

$$f(x) = \int e^{i\lambda} d\mu(\lambda)$$

in the usual notation of the real line or the circle. To be more precise, a measure "on" a set is a measure whose support lies inside this set, that is, it is null outside it. We say that the set $S$ is a set of spectral synthesis if all functions whose spectrum lies inside $S$ can be reconstructed from measures whose support is in the closure $\overline{S}$ of $S$. We say that spectral synthesis holds on $L^1(G)$ if every such set is a set of spectral synthesis. We say that spectral synthesis fails on $L^1(G)$ if there are sets which are not sets of spectral synthesis: To reconstruct a function whose spectrum lies inside such a set, we need **additional components** from outside its closure. The theorem of Malliavin states that for the case of a commutative, locally compact, **non compact** group, spectral synthesis fails on $L^1(G)$.

# CHAPTER 3

# EMPIRICAL DISTRIBUTION FUNCTION

We now turn to the discussion of the first one of the methods that may be used to estimate a probability density. It is known in the literature as the empirical distribution function. This method estimates the distribution function of a random variable given the experimental data, and the estimate is expressed directly in terms of these data, not in terms of the moments as is proposed in this thesis, but it will be discussed here for the sake of completeness and to be able to show the merits of the core approach in this thesis. It will have its place in the full picture when the various methods presented are compared in a later chapter.

## 3.1. Background

The analysis will be carried out following [Rao], [Lev] and [Gne]. An important contribution to this method appears in the paper of Parzen, [Par], where the author presents a method to construct a density from the empirical distribution function (It will be referred to as the edf). The asymptotic formulae and estimates are brought from [Lev] and [Gne], and the proof of the main convergence theorem of the edf can be found in [Rao]. The paper [Mas1] treats a method of p.d.f. estimation based on the density estimation of Parzen. Finally, important results in [Ahe] and [SohT] will be cited, mainly concerning the relation between moments and experimental data. These two later references, as well as references cited therein, offer a mathematical view point of the "moment problem" which is rich in results, and these results deserve a complete overview to assess their importance. The cited results were included in this chapter because the edf is proposed (in the references) as a solution to the moment problem, although this problem is phrased in terms of moments.

## 3.2. Exposition of the method

To begin the exposition, consider an observed sequence $(x_i)_{1 \leq i \leq n}$, which we consider as a sequence of trials for a random variable $X$, of unknown distribution (In fact, it is the distribution we want to find) $F(x) = P\{X \leq x\}$. Each item of data $x_i$ is a realisation of a r.v. $X_i$, these constituting a sequence of independent identically distributed (i.i.d) random variables, of common distribution $F(x)$. The problem is to find, or more precisely, estimate, this distribution, given the observations. The edf corresponding to these observations depends on the number $n$ and is defined to be

$$F_n(x) = \frac{1}{n}\{\text{number of } X_i < x\} \qquad (3.2.1)$$

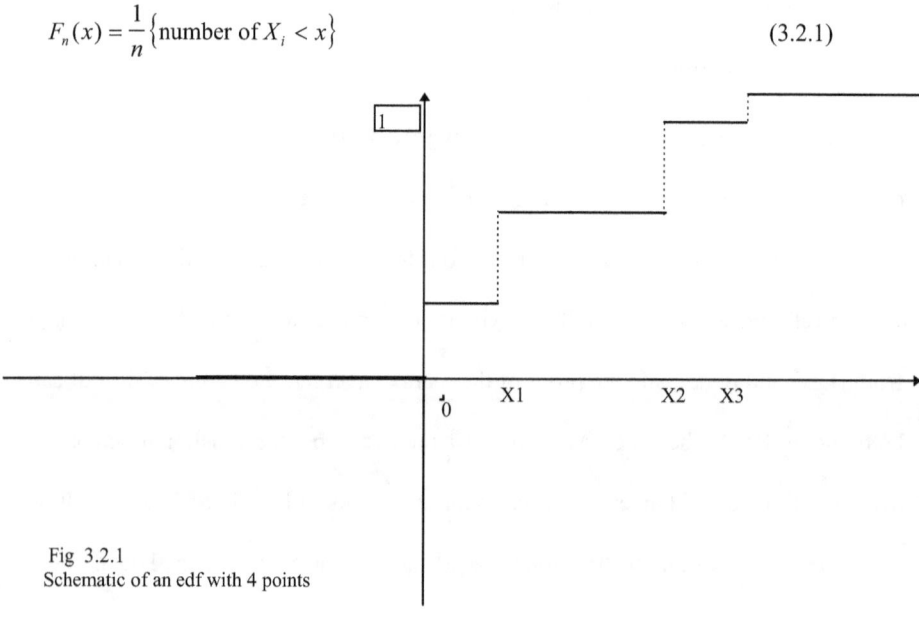

Fig 3.2.1
Schematic of an edf with 4 points

To better understand and evaluate this definition, let's compare it with the general definition of a distribution function $D(x)$, corresponding to a r.v. $R$ :

$$D(x) = \text{Prob}\{R(\omega) < x\} \qquad (3.2.2)$$

We see that the probability of an interval on the real line, in the measure defined by the distribution we are to construct, is the relative number of items of data falling in that interval: The edf assigns to an item a probability proportional to its frequency of occurrence in the sequence of observations. The p.d.f. associated with this distribution may shed more light on this approach. This p.d.f. is the derivative of $F_n(x)$, and it consists of a sequence of impulses located at the data points $x_i$:

$$p_n(x) = \frac{dF_n}{dx} = \frac{1}{n}\sum_{i=1}^{n}\delta(x - x_i) \qquad (3.2.3)$$

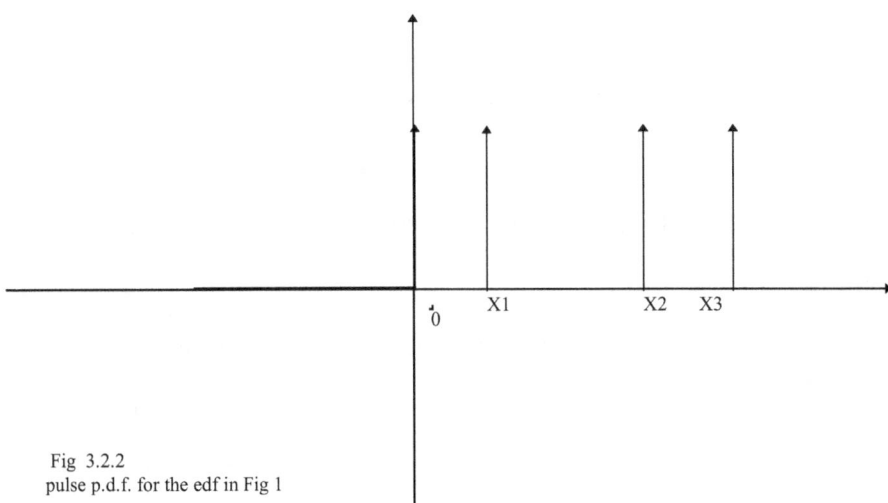

Fig 3.2.2
pulse p.d.f. for the edf in Fig 1

This strict localisation at a discrete set of points is in conflict with the basic characteristic of a p.d.f., seen according to the point of view held in this thesis: That a p.d.f. represents a **pattern** of data. In a certain sense, it is inconsistent with its supposed nature. In fact, we can see that it is discontinuous, when looked at in the context adopted thus far of a continuous (as opposed to discrete) density.

Since the mentioned inconsistency results from the fact that we rely on a discrete set of data to perform the estimation, that is, the estimate is derived **directly** from this discrete set, the natural starting point is to try to tackle this "discreteness". There are two alternatives approaches to perform this task: to operate on the data first, or to operate on the distribution function or the density. If we look at the data, considering the observations as discrete amounts practically to consider them as "few", not numerous, and one could think of adding more data in some way. On the other hand, looking at the density, which is a sequence of discrete pulses, or at the edf, which has discontinuous jumps at the data points, the obvious answer is to use a smoothing function. Let's look first at the first alternative. As we have said, it follows from a judgement of the observer that the amount of data observed is not enough. Since the very aim of observation is to draw a conclusion, or to make a decision, the observer is faced with another set of possibilities. One is that the process which generated the data is available for more observations, a possibility which spontaneously drives us outside of the actual alternative (not enough data). The next possibility is to try to generate "hypothetical" data according to a certain procedure, which must be derived from the data itself. We can expect here the appearance of a role for the moments, since these, in fact, describe the pattern of data, and it can be shown, as a consequence of a theorem in [Ahe] (to be stated later in this chapter), that there may be many configurations of data that share the same pattern, that is, the same moments. This role of the moments will appear persistently throughout the steps of analysis of the edf method, a fact which will put in evidence the superiority of the moments approach. The final possibility is to change the "scale" of observation, in a way which will be described shortly in formal and analytical terms. But the outcome of this scaling will be to drive the observer another time outside the scope of the first alternative which is still discussed now. The idea is to regard the set of observations, considered as "not large enough" as a subprocess of a wider scope process, and to be content to stay inside the scope of the newly defined subprocess. In this situation, the data are enough to describe this later process, and we turn now to the second alternative, which is supposed to operate at the level of the distribution or its density.

It will be shown that this alternative describes the case of "large or relatively large set of observations".

The idea is that, considering the number $n$ of observations $(x_i)_{1 \le i \le n}$ as a large set is expressed mathematically as looking at the "behaviour" as $n$ approaches infinity, $n \to \infty$. This "behaviour" is evidently described, at least at the level of this work and in its context, by the distribution function or the density governing the observed data. The result is that the assertion "there's a large enough set of data" leads naturally to the study of the convergence properties of the edf $F_n(x)$ for $n \to \infty$. Rigorously, since $F_n(x)$ is a function of the sequence of i.i.d random variables, the $(X_i)_{1 \le i \le n}$, it is itself a random variable, and the problem of convergence should be studied in the context of the theory of convergence of sequences of r.vs. . We have the following result from this theory of convergence [Rao]:

Theorem 1: Given a sequence of independent identically distributed random variables $X_i, i = 1, 2, \cdots$, obeying the same distribution $F(x)$, the sequence of distributions $F_n(x)$ given by equation (3.2.1) converges **in probability** to the common distribution $F(x)$, as $n \to \infty$:

$$\Pr ob \left\{ \lim_{n \to \infty} \sup_{-\infty < x < \infty} \left| F_n(x) - F(x) \right| = 0 \right\} = 1 \tag{3.2.4}$$

Now recall that, according to the second alternative, a certain operation of smoothing of the edf or the derived impulse density is required. The theorem just stated, however, doesn't explicitly show any kind of smoothing, but we expect the common and limit distribution, $F(x)$, will be a smooth function in general, whereas the elements of the sequence, the $F_n(x)$, still are discontinuous, even for large $n$. These functions have to be rendered continuous in order to be able to provide general results and to be tackled analytically in practical problems. This question is answered, at least partially, by a method due to Parzen, in the paper [Par], which will be discussed shortly. But first, we pause a little to look at the results obtained thus far in the light of the main theme of this thesis, the moments.

The first appearance of a possible role for the moments in the context of the edf method in this chapter has been in the discussion of various possibilities that may arise in the "first alternative", in which the observer asserts that there's not enough data. In this situation, the moments would be one possible way to provide a **pattern** for the actual observed data, thus ensuring the existence of other hypothetical data that are not foreign to the data in hand. This is an intuitive result that will be supported rigorously, as has been mentioned at the occasion, by a result from [Ahe], stating that there's an infinity of possibilities for sets of data, given a certain finite sequence of moments. Now if we generate a certain amount of additional, hypothetical data so as to be out of the first alternative and into the second one (enough data), and proceed to the smoothing which will be discussed shortly, the result should not be far from estimates based on the moments approach of this thesis, since then the data available will share a highly stable pattern given by the set of moments. But the main question is to decide as to when should the observer considers that the data set is large and when is it a small set. The common distribution is not known, since it is the one we're looking for; however, it is only according to this distribution that it is possible to tell, in a reliable way, which alternative to chose. Without the moments, this choice will be either arbitrary or based on "engineering judgement" of an experienced observer. Even if this experienced judgement is considered to be highly reliable in one or a few situations, it can not be so for a whole set of different situations. Moreover, it will be difficult to quantify its performance level and the errors that may arise from it. As the discussions in chapter one has shown, all these problems are tackled by the moments approach. The arbitrariness is concentrated in only one point, which is the choice of moments order, a problem that can be cast in a general way encompassing many situations, and its performance or drawbacks can be quantified analytically, being the order of the Taylor expansion of a certain function. Informally stated, the use of moments is a **conscious procedure**, in contrast with the experienced judgement, which is simply a good **guess**. This aspect of the difference between edf and moment approach will appear at every step in the edf method. It will appear when discussing the scale parameter choice and smoothing function choice, which we turn to now, in order to continue the main theme of this chapter.

Recalling the discussion of the two alternatives for obtaining the required distribution, the first choice was to generate hypothetical, or, if possible, real experimental data, to reach the situation where the data set is large enough to be a valid representation of the pattern that should be described by the solution distribution. Theoretically, there's no way to be sure of the point where this process of data filling has to stop, since this point depends on the sought solution itself. The only way out of this loop is to make a leap to the continuum and smooth the discrete density to obtain a continuous one. There are two main aspects, or parameters, the smoothing function should deal with. One is the behaviour of this function for large values of the variable. The answer is that, since it is supposed that the point has been reached where the available data provide a genuine pattern of the true situation, then the smoothing function should decay fairly rapidly beyond the final items at the boundaries of the set of data. The other point this function has to deal with is the behaviour in the gaps between the data items. In the method proposed by Parzen in his paper [Parzen], this aspect is dealt with separately from the smoothing function, as an additional "scale" parameter or set of parameters. The reason is clear, the mentioned gaps vary randomly throughout the data, they are numerous, and many possible sets of data representing the same pattern have to be addressed uniformly, in a single way. Let's state now the proposed solution to make these points clear. In a simple form with only one scale parameter, we have

$$p_c(x) = \int_{-\infty}^{\infty} K(\frac{x-t}{h}) p(t) dt \qquad (3.2.5)$$

where $p_c$ is the continuous, solution p.d.f., $p$ is the original, pulse p.d.f., $K(u)$ is the smoothing kernel, and $h$ is the mentioned scale parameter. The result is

$$p_c(x) = \frac{1}{nh} \sum_{i=1}^{n} K(\frac{x-x_i}{h}) \qquad (3.2.6)$$

The requirements for the kernel are [Rao]

$$0 < K(y) < \infty, \quad \lim_{y \to \pm\infty} yK(y) = 0,$$
$$\int_{-\infty}^{\infty} K(y) dy = 1 \qquad (3.2.7)$$

As examples of such a function, we can mention two candidates:

$$K(x) = \frac{1}{\pi} \frac{1}{1+x^2}$$

and

$$K(x) = \frac{1}{\sqrt{\pi}} \exp(-x^2)$$

Let's look at the behaviour of the new p.d.f. in the neighbourhood of one of the $x_i$. If $h$ is too small, and we consider a point $x$ not so close to $x_i$, then the term $\frac{1}{h} K(\frac{x-x_i}{h})$ behaves, for $h \to 0$ ($h$ too small), like $yK(y), y \to \pm\infty$, where $y = \frac{1}{h}$, and therefore this term, and all the other terms in the p.d.f. will go to zero, in consequence of the conditions (3.2.7) prescribed on the kernel, so that this value of the variable $x$ is considered as an invalid data item. On the other hand, keeping the same condition on $h$, $h \to 0$, and considering $x - x_i \approx h$, the argument $\dfrac{x - x_i}{h}$ is finite and the value of the p.d.f. will be given by a finite (not too small) number divided by an infinitesimally small $h$, the outcome being a large number. The interpretation of these remarks is straightforward. The operation of filling the gaps is performed "near" the experimental, actual data. $h$ serves as an indicator to how much some point will be near these real data. In other words, $h$ sets the **scale** for the empirical process. In the situation just described, $h \to 0$, it is only on an interval of the size of $h$ that the data are considered to be packed densely. We can expect that for the case of a large set of data (the second alternative), the size of the scale parameter $h$ must be chosen to be comparable with the smallest gap between data in the pulse density, that is

$$h \approx \inf_{k} \left| x_k - x_{k+1} \right| \tag{3.2.8}$$

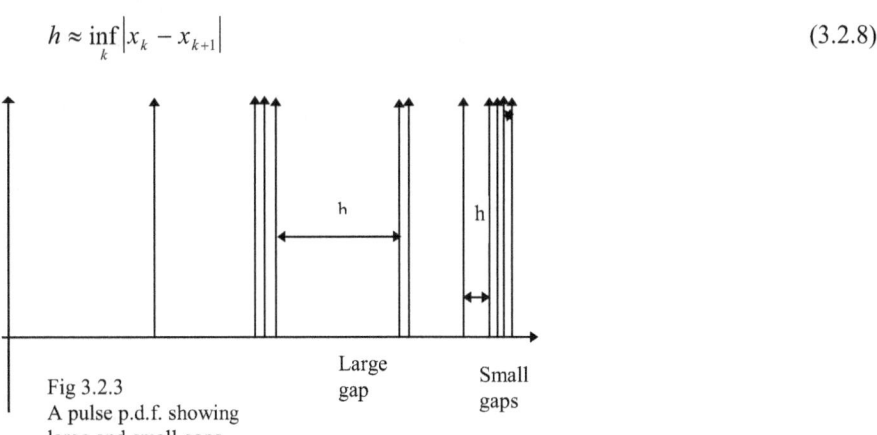

Fig 3.2.3
A pulse p.d.f. showing
large and small gaps

That's because the data is supposed to represent faithfully the general pattern given by the required p.d.f.. For gaps large with respect with this chosen value for $h$, we'll have a small value of the smoothed density, a fact consistent with this latter assumption: Adding too much data in such a gap would mean that we have possibilities not represented by the actual data. In contrast, in intervals where the data is densely packed , which means a gap of the size of the chosen $h$, we'll have a large value for the p.d.f., in accordance with the expectation given by the dense set. Recall now the different possibilities for the first alternative (not enough data), where one of them was to change the "scale" of observation, and to consider the observed data as representing a subprocess of the original, whole process. For this subprocess, the data would be representative, and it is possible to treat this case in the same way as for the case of enough data, and it is proposed that this can be done by choosing a larger scale parameter, i.e. a larger value for $h$. Then we'll have that over intervals of length comparable to this $h$, the data are representative, and the value of the estimated (smoothed ) p.d.f. will be finite, not too small, over a larger range of intervals.

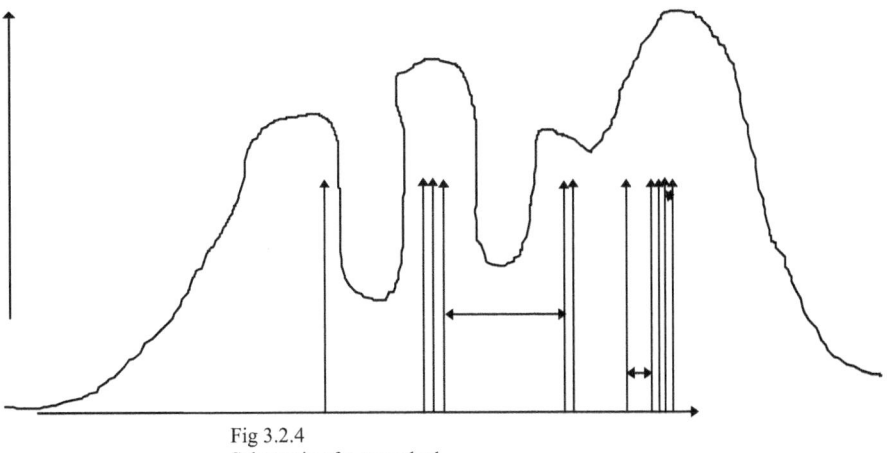

Fig 3.2.4
Schematic of a smoothed
p.d.f. with small h

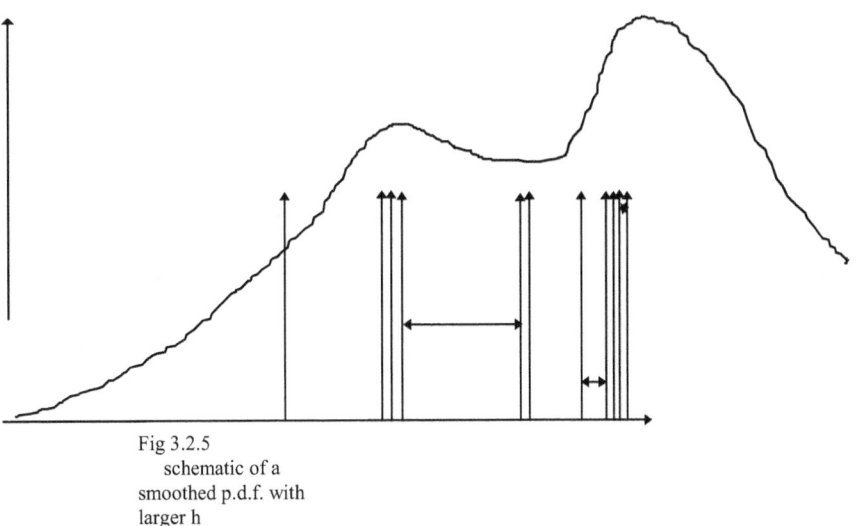

Fig 3.2.5
schematic of a
smoothed p.d.f. with
larger h

To close this section, we state the convergence theorem for the smoothed

density. We have[Rao]:

Theorem 2:Given a sequence of i.i.d random variables, with common

distribution $F(x)$ which has a density defined on $\mathbb{R}$, $f(x) = \dfrac{dF(x)}{dx}$, and a sequence of real numbers $h(n) > 0, \lim_{n \to \infty} h(n) = 0, n(h(n))^2 = \infty$, then the sequence of smoothed

p.d.fs. given in (3.2.6)

$$f_n(x) = \frac{1}{nh(n)} \sum_{i=1}^{n} K(\frac{x - x_i}{h(n)})$$
(3.2.9)

where $K(u)$ satisfies the conditions (3.27), converges in probability to the common

density $f(x)$

$$Prob \left\{ \lim_{n \to \infty} \sup_{-\infty < x < \infty} |f_n(x) - f(x)| = 0 \right\} = 1$$
(3.2.10)

## 3.3. Asymptotic behaviour

In the previous section, two convergence theorems were stated showing that the proposed edf its density estimates were effectively valid estimates for $n$, the size of the observed data set, sufficiently large. However, in practical situations analysis, and particularly in consideration to the main objective of this work, this is not enough. The question of the **rate of convergence**, or equivalently the **approximation error** committed when stopping at some stage represented by some large value of the number $n$, have to be addressed. We'll review here some asymptotic formula and try to clarify the processes of convergence in theorems 1 and 2.

The first result is due to Kolmogorov [Lev], [Gne], who derived a formula giving the rate of convergence of the error random variable defined by

$$\Delta_n = \sup_{-\infty < x < \infty} |F_n(x) - F(x)|$$
(3.3.1)

The result is that the distribution of the quantity $\Delta_n \sqrt{n}$ follows asymptotically the law

$$\Pr ob\left\{\Delta_n \sqrt{n} \le z\right\} \approx \sum_{k=-\infty}^{\infty}(-1)^k \exp(-2k^2 z^2) \qquad (3.3.2)$$

Which can be put in another way because of the symmetry of the term $k^2$:

$$1 + 2\sum_{k=1}^{\infty}(-1)^k \exp(-2k^2 z^2) \qquad (3.3.2')$$

Let's carry out a simplified analysis of this asymptotic behaviour. For $z$ large, it is possible to restrict the expansion to a few terms, since the term $\exp(-2k^2 z^2)$ in this situation becomes rapidly too small for the first values of $k$. For example, we may have

$$\Pr ob\left\{\Delta_n \sqrt{n} \le z\right\} \approx 1 - 2\exp(-2z^2)$$

The term on the right hand side is close to one for $z$ large, a conclusion consistent with what we would expect for any distribution function: The quantity $\Delta_n \sqrt{n}$ is surely smaller than a large value. However, this estimate quantifies, to a certain degree, the qualitative behaviour expected from the distribution. The consistency remains observed for small values of $z$. In that case, there's a relatively low probability for $\Delta_n \sqrt{n}$ to be too small, and the exponentials are not negligible and have a substantial value close to one, resulting in a low value for the whole expression.

The quantity $\Delta_n$, as defined above, is one possible norm measuring the deviation of the edf from the true distribution. Another possible norm is the weighted square norm, where we define the quantity

$$D_n = \int_{-\infty}^{\infty}\left|F_n(x) - F(x)\right|^2 f(x)dx \qquad (3.3.3)$$

the function $f(x)$ being the p.d.f. corresponding to the limit distribution

$$f(x) = \frac{dF(x)}{dx}$$

The role of this weighting density is to emphasise the difference error at points of high probability ($f(x)$ large), and de-emphasise this error at unimportant points ($f(x)$ small). This error measure is attributed to Von Mises in [Lev], where its asymptotic behaviour is calculated to be

$$D_n \approx \frac{1}{12n^2} + \frac{1}{n}\sum_{k=1}^{n}\left[F(x_k) - \frac{2k-1}{2n}\right]^2 \qquad (3.3.4)$$

where the $x_k$ are the observed data. We can see that this estimate depends also on the limit distribution $F(x)$, more specifically on its values at the points $x_k$.

Again, we remark that the use of these estimates in practical situations will rely on subjective judgement and experience of the observer. An estimate telling that some error is "close" to some function of $n$ for "large" $n$ doesn't specify how close is it, and how large $n$ should be. As has been said before in the previous section of this chapter, using the moments would provide a solution to this problem by concentrating the arbitrariness on one point only: The choice of moment order. These points will be fully analysed in chapter 6, where the various methods are compared, but now we anticipate this analysis by looking at the moments of the edf and the density estimate using the smoothing kernel.

## 3.4. Moments of the edf

Let's calculate the moments of the edf by integrating the nonsmoothed, pulse density given by (3.2.3)

$$p_n(x) = \frac{1}{n}\sum_{i=1}^{n}\delta(x - x_i) \qquad (3.2.3)$$

We have

$$m_k = \int x^k p(x)dx = \frac{1}{n}\sum_{i=1}^{n}\int x^k \delta(x - x_i) = \frac{1}{n}\sum_{i=1}^{n} x_i^k \qquad (3.4.1)$$

We see the expected result that the moments of the "raw", nonsmoothed density, or equivalently, of the edf, are the experimental moments deduced from the observations. An important point here is that the act of considering these moments to **all** orders, without any operation on them, amounts to ignoring their role at all. It can be said that using the edf as an estimate to get information corresponds to "guessing", without any intellectual "thinking", whereas this latter is represented by the use of moments. In other words, as we have said previously, the moments represent a conscious procedure following determined steps (the higher order moments).

Second now, we look at the moments of the smoothed p.d.f.. These are given by

$$m_{1k} = \frac{1}{nh}\sum_{i=1}^{n}\int x^k K(\frac{x - x_i}{h})dx \qquad (3.4.2)$$

After making the change of variables $u = \dfrac{x - x_i}{h}$ and performing the integrations and summation, we obtain

$$m_{1k} = \sum_{i=0}^{k} C_i^k h^i r_i m_{k-i} \quad , \quad k = 0,1,2,\cdots \qquad (3.4.3)$$

In this equation, the $r_i$ are the moments of the kernel $K$, considered as a probability density, the $C_i^k$ are the known combinatorial coefficients and arise from the binomial expansion of the term $x^k = (hu + x_i)^k$, and $m_0 = (m_1)_0 = r_0 = 1$. We can see readily an important feature of this equation: The new moments are obtained from the moments of the kernel $K$ through a filtering operation, specifically a linear "time **variant**" operation. The coefficients of the filter are the $C_i^k h^i m_{k-i}$, which depend on the "time" $k$ through the terms $C_i^k$ and the experimental moments $m_{k-i}$.

This remark may suggest the possibility of applying methods in filtering to the sequences of moments we have here. More specifically, this application may be that the "error" given by $m_{1k} - m_k$, taken as a function of the "input" sequence $(m_k)_{1 \le k \le n}$, is to be minimised with respect to the parameters of the filter, the $(r_i)_{1 \le i \le n}$, so that the indeterminacy in the choice of the scale parameter $h$ and the kernel $K$ is resolved by solving the minimisation problem. The first thing to say about this idea is that, if it can be implemented in some way, it will lead to the problem of estimation of a density $(K)$ from its moments $((r_i)_{0 \le i \le k})$, and we are brought into the main theme of the thesis. However, a close examination of the equation(3.4.3) will show that this is an unfeasible problem. In fact, in analogy with filtering theory, we can think of finding the least squares method. In this context, we consider the error vector

$$\underline{\varepsilon} = (\varepsilon_l)_{1 \le l \le k} \quad , \quad \varepsilon_l = m_{1l} - m_l$$

From equation (3.4.3), this quantity is given, in terms of the $(r_i)_{1 \le i \le k}$, by

$$\underline{\varepsilon} = A.\underline{r} \tag{3.4.4}$$

where $\underline{r}$ is the moments vector whose components are the $(r_i)_{1 \le i \le k}$, and $A$ is a square matrix given by

$$A_{li} = C_i^l h^i m_{l-i} \tag{3.4.5}$$

The minimisation problem is of finding the minimum of

$$\|\varepsilon\|^2 = \underline{r}^T A^T A \underline{r} = \underline{r}^T Q \underline{r} \quad , \quad Q = A^T A$$

The derivative of this expression is $2Q\underline{r}$, and can't be equal to zero unless the matrix $Q$ is singular, and thus so would be the matrix $A$. From equation (3.4.5), we can infer that the determinant of $A$ is a homogeneous polynomial in $h$, since column $i$ is a multiple of $h^i$ for every $i$. And so, apart from the trivial solution $h = 0$, there are solutions for very specific cases for specific values of the experimental moments

$(m_l)$. All of what has been said can be rephrased in terms of filtering theory by saying that the filter given by the equations (3.4.3) is unrealisable because it is not causal: The output at "time" $k$, which is $m_{1k}$, depends on the input at the same time $k$, $m_k$.

## 3.5. A result from the mathematical theory of moments

We cite in this section a main result from [Ahe] concerning the reconstruction of data from a sequence of moments. [SohT] is a similar reference, and both of these works treat the mathematical moment problem using tools from real and complex variable functions theory. It comes clear in these works that this problem has been an important theoretical question, and has connections with other fields of mathematics. We follow here mainly [Ahe].

We begin by considering a sequence of numbers $(m_k)_{0 \le k \le n}$. This sequence is supposed to be positive definite, in the sense that the matrix $M = (M_{ij})_{0 \le i,j \le p}$ , $p = \left[\dfrac{n}{2}\right]$ (where the brackets indicate that the integer part of the number within is considered), given by $M_{ij} = m_{i+j}$, is positive definite. This matrix will be used in the computation of the coefficients of the maximum entropy solution, and the fact that it is positive definite will appear to be an important feature. Now the following theorem shows the connection between a sequence of moments and the data that may generate it. We have [Ahe]

<u>Theorem 3</u>    a)  If a sequence $(m_k)_{0 \le k \le 2p-2}$ is positive definite. There exist infinitely many canonical representations

$$m_k = \sum_{i=1}^{p} \rho_i \xi_i^k \quad , \quad k = 0,1,\cdots,2p-2 \tag{3.5.1}$$

where

$$\rho_j > 0 \ , \quad -\infty < \xi_1 < \xi_2 < \cdots < \xi_p < \infty$$

b) The systems of numbers $(\xi_i)_{1 \leq i \leq p}$ in two distinct representations alternate.

c) Among the representations there exists one and only one representation in which the system of numbers $(\xi_i)_{1 \leq i \leq p}$ contains an arbitrary given point $\eta$ (The arbitrariness is subject to a specific restriction not mentioned here).

d) Given any real number $m_{2p-1}$, among the representations we can find one and only one such that

$$m_{p-1} = \sum_{i=1}^{p} \rho_i \xi_i^{2p-1} \tag{3.5.1'}$$

Let's analyse these statements in the light of what has been discussed so far. The first thing to note is the significance of the coefficients $\rho_i$. We can see that, since these coefficients multiply the "data" $\xi_i$, they should represent, and they look like the discrete probabilities of these data. In fact, the expression (3.5.1) corresponds to a discrete state system (or random variable), the probability of each one of these states, the $\xi_i$, being equal to $\rho_i$. So the foregoing theorem 3 gives, in some way, the solution of the problem of estimation of a p.d.f., but as a discrete state density. And this density is not unique (nor is the set of states), since the theorem asserts the existence of an infinite number of canonical representations, or solutions. On the other hand, speaking in terms of the edf, the $\rho_i$ will represent the "number of occurrences" of the $\xi_i$, although this statement is not quite accurate, since the $\rho_i$ are not necessarily integers. In case they are, we will obtain a set of data from which the moments may arise. It is not clear whether a representation with such integer coefficients is unique, but we can say that, "around" some arbitrary one of the representations, we can adjust the $\rho_i$ to be integers, eventually creating (or destroying) data items $\xi$, to obtain approximately valid representations, which are apt to be considered as edf candidates. Since this is performed on an approximation base, we can, first conceptually, think of merging these discrete representations into one, or may be more, continuous solution or solutions which are continuous. We expect these solutions to correspond in some way to the smoothed version of theedf, with eventually different solutions emanating from different smoothing kernels. In all cases, however, we can say that, at least approximately, for one sequence of moments there correspond an infinite set of data that may generate this sequence. In this respect, we note that for the particular case where the number of data items looked for is equal to the number of moments (the moment order considered or available), we obtain an algebraic set of equations with equal number of equations and unknowns, the equations being polynomials in the unknowns

$$m_k = \frac{1}{n}\sum_{i=1}^{n} x_i^k \quad , \quad k = 1,2,\cdots,n$$

The theory of polynomial algebraic equations doesn't forbid multiple solutions for the system of equations in case of more than one variable. However, the set of solutions must be discrete and finite.

# CHAPTER 4

# LEAST SQUARES METHOD

With this chapter we begin the core of the thesis, where we start the exposition of the least squares method, or what we have called the context method, the minimum distance method and the semantics approach to the problem. Two key features of this method are the square norm (In fact, as a norm, it is the square root norm) and the weighting function. The discussion in this chapter will begin by a review of the background for the method underlying these two features, and so we will need first a brief enunciation of the form of the problem to give an idea of where these concepts of least squares and weighting function do appear, and in what formal context.

First the weighting function. In this regard, recall the context of appearance of the characteristic function in the two basic identities

$$\Phi(\lambda) = \int \exp(i\lambda x) p(x) dx$$

and

$$\Phi(\lambda) = \sum_{k=0}^{\infty} \frac{i^k m_k}{k!} \lambda^k$$

Now, having the moments up to a certain finite order $N$, $(m_k)_{0 \leq k \leq N}$, we form a truncated characteristic function

$$\Phi_T(\lambda) = \sum_{k=0}^{N} \frac{i^k m_k}{k!} \lambda^k$$

This truncated function is a polynomial in the variable $\lambda$, and so diverges at infinity. More precisely, It approximates $\Phi(\lambda)$ well over a limited interval, outside of which it will begin to diverge, and the more higher the moments order is, the more larger is this interval. Ultimately, going to an infinite order will yield again $\Phi(\lambda)$, which is assumed to have a full Taylor expansion and to decay to zero at infinity. It doesn't have an inverse Fourier transform, and it is not possible to recover a probability density from it (in the usual sense, because it is possible in the sense of distributions). Here appears the role of the weighting function. This function is used to multiply the truncated characteristic function to force it to be well behaved at infinity, making it convergent to zero, so that the integration to form the inverse Fourier transform can be carried out. The new obtained function on which operations can be now performed is $W(\lambda)\Phi_T(\lambda)$, the first one, $W(\lambda)$ being the weighting function.

The least squares approach is now to find the closest function to $W(\lambda)\Phi_T(\lambda)$ in the least squares sense. We proceed now to the background of this formulation.

## 4.1. Background

A major part of the background on the theme of this chapter is spread over a wide variety of domains and works in the literature, so some of this background is in a certain sense "contextual", where an idea appears in many contexts and applications and in different texts and research works. Only typical references are mentioned in these situations.

The square root norm has its roots in the ordinary distance in classical geometry, or, in modern terminology, in a Euclidean space. In classical geometry, it is expressed in the form of Pythagorus theorem on the length of the hypotenuse in a rectangle triangle; in the latter context, that of modern Euclidean space, it is expressed as the quadratic form derived from a positive definite symmetric bilinear form. But its use in more abstract sets as a measure of "error", which is its usual common use in the field of signal processing and control systems engineering, may seem a bit unjustifiable at the fundamental level in spite of the success it achieved in applications in these fields. As a criterion in optimisation problems, it has the particularity and advantage of leading to linear equations, in contrast to other possible kinds of criteria. Now in the approach followed in this thesis, this type of justification is exactly the one we're trying to avoid: The property of leading to easily (relatively) solvable equations. A more fundamental objection is that this criterion treats abstract spaces, consisting in extremely diverse types of mathematical objects (the diversity meant is between elements of one space and elements of another space) as if they were points in ordinary geometric space, so that a point of such an abstract space is similar to a "location" in our ordinary, usual space, something which is not obviously true at the first sight. This kind of use of this norm dates back to the German mathematician Gauss (end of the eighteenth century), who used a least squares method to calculate the orbits of planets and asteroids. The reference is the paper by Gauss himself [Gau]. According to him, the unknown parameters of the mathematical model should be chosen in such a way that (the quotation is from [Ast])

"the sum of the squares of the differences between the actually observed

and the computed values, multiplied by numbers that measure the degree

of precision, is a minimum."

This statement contains also the idea of the weighting function: "the numbers that measure the degree of precision". These numbers multiplying the different components of the vector error give different "weights", and therefore different degrees of emphasis on each of these different components. After Gauss, the French mathematician Legendre published a paper [Leg] treating the least squares method. Subsequent works recorded the use of this method specifically in areas of estimation theory, a topic which at the time of the two mentioned papers was almost unknown except from rudimentary concepts of probability theory, will wait until the thirties and forties of this century to appear. Before proceeding to a brief review of these works, some other kind of background will be discussed, which we may call logical or contextual, in the sense that it relates to the mathematical foundations of this least squares method, and which are present in a variety of mathematical subjects.

The discussion of the context method concept in the first chapter cleared the fundamental objection on the least squares approach: Objects are understood only in a certain specific context, and their "understanding" is represented by a scalar product which reduces pairs and multiplets of entities to more readily assimilable products, the scalars (usually real numbers). The ordinary space is simply a specific instance of context, or may be considered as an abstraction of this concept, then providing the observer, in the same time, with the basics of the whole idea (of context). One basic foundation, or may be **the** basic foundation of the mathematical square root norm is in the concept of duality. At a more technical level, we find the well defined tool of Hilbert spaces, based on the scalar product, where the concepts of orthogonality and expansion of an element in terms of elements of a complete orthonormal system appear. According to the discussion in the first chapter, the scalar product is some form, or some outcome of the concept of duality. This hypothesis is strongly supported by the introduction, in the theory of Hilbert spaces, of the **dual** space, which is the space of continuous linear functionals on the original space. This picture is completed by a well known result: Taking the dual of the dual space, we recover the original one. This concept of duality finds similar role in varied subjects of mathematics, such as group theory, and especially in harmonic analysis, whose subject is the study of the structure of topological groups, and this is again a strong support for the ideas presented in this thesis about these concepts. We can find ramifications in number theory and algebraic topology (Homology and Cohomology). Finally, at a point of intersection between such abstract developments and the practical world of signal processing and control theory, we find topics such as orthogonal polynomials and expansions in terms of such polynomials, which are particular cases in harmonic analysis and Hilbert space theory; and we can mention the famous Karhunen-Loeve expansions as a very specific example of interplay between this theory and random processes analysis.

The beginning of properly said "least squares" methods in the fields of signal processing and control systems theory dates back to fundamental works of Kolmogorov and Wiener. The brief review here derives from [Kailath], where are reproduced main papers which are milestones in the progress of the least squares estimation, including the paper [Kai2] which reviews completely this field in preceding years, back to the thirties. In his paper [Kol2], Kolmogorov was the first to formulate the problem of least squares prediction and extrapolation of stationary time series. Wiener, in [Wie], solved this problem and obtained the well known Wiener-Hopf equation. In another paper [Kol3], the author formulated estimation problems in terms of the Hilbert space theory. Another main development is the introduction of state space methods in signal processing, which appeared in the work of Kalman and Bucy [KalB]. The filter so obtained, known as the Kalman filter, gives the optimum estimate of the state in the least squares sense. It is also shown in this same paper that the estimation problem is dual to the optimum regulator problem (optimum control), optimal in the least squares sense.

An important point, relevant to this background review, is the inherent characteristic of the least squares problems to yield decomposition of sets (linear spaces or subspaces) into smaller ones, a fact expressed in a form common in all vector spaces as the expansion in terms of the elements of a base (or a complete system, for Hilbert spaces). In the context of this discussion, the Karhunen-Loeve expansion (the paper [Kar], [Loe], [Tre])of a continuous time random process in terms of a countable, discrete set of orthogonal functions, is a well known example. This approach has been used in the discrete, finite dimensional case, for the eigenvalue decomposition of the covariance matrix of a random time series, in what is known as the Karhunen-Loeve transform [AkaH]. The strong connection will appear shortly, while turning to the discussion of the many uses of the "weighting" process, thus setting the background of the tool of "weighting function" proposed in this thesis. In fact, in mathematical terms, a least squares estimation is an operation of projection on a certain subspace, containing "admissible" elements, the solution searched being supposed to be in this subspace. Thus the observation is supposed to be composed of two components, and the observer is interested in one of them, and so he will give minimum interest in the second component, which is exactly what is meant in trying to minimise the error (least square error). Now, in general, the error space may be broken down into smaller subspaces: It is a multidimensional vector. And the observer can perform the weighting on these components so that, while not completely discarding some of them, he may well give them smaller "weight", that is, smaller coefficients in the expression of the least squares criterion. It can be seen clearly that we have reached naturally the point where an element in a vector space (Hilbert space) is decomposed in terms of the elements of a basis, which is an intrinsic property of linear spaces. This type of reasoning is justified in the light of the discussion in chapter one of the context concept. What we have done now is that the "degree of importance" or "weight" concepts have been **called** many times, **recursively**. This side of the least squares problem will be shown to be a particular characteristic of the variant of the method used in this chapter.

The transition from "least squares" to "weighting" having been achieved, we discuss here another contextual or logical background for the weighting function idea proposed to be a main ingredient in the formulation of the main theme of this chapter. The first type of context, and the basic one, is the ordinary and common one just discussed: In an estimation problem where we seek to minimise the square norm of the error vector

$$e = (e_1,\ldots,e_n)^T$$

to obtain the criterion

$$e^T e = \sum_{i=1}^{n} e_i^2 \rightarrow \min$$

Now this is an expression corresponding to a particular quadratic form, and nothing forbids us from using just another one, which can be expressed in the same basis as for the previous equation as

$$e^T Q e$$

where $Q$ is a positive definite symmetric matrix. The first expression corresponds to the case where $Q$ is the identity matrix. Now to recover a weighted expression, we diagonalise $Q$ (an operation which is always possible since we are dealing with a positive definite symmetric matrix; this is an elementary fact in linear algebra) to obtain the second expression in the new form

$$\sum_{i=1}^{n} \lambda_i e_i^2$$

and we obtain a weighted sum of the squares of the components of the error vector. The physical interpretation of this operation is that highly weighted components are important, while the ones with lower weights are not as important, and their contribution to the total value of the optimality criterion is de-emphasised. We have mentioned at the beginning of this chapter that Gauss used such weighted error in a parameter estimation problem. This case is treated in standard texts on estimation theory, and here we can mention [Ast], where a well known expression of the parameter vector estimate is obtained for the more general case of a matrix different from the identity matrix.

Now the criterion of an optimisation problem doesn't have to be always as a minimal error criterion. A more general setting is found in the theory of optimal control, where the controller part of a system is required to achieve some "cost" function minimisation. Such a function may be the time taken by a moving object to reach a certain specified target, the amount of fuel consumption in such a case, **or a combination of both**. It may be the combined goal of maximum yield of chemical reaction and a minimum temperature or pressure. This can be called a multicriteria control, and it is clear that we need to specify the importance of each single component of the combined criterion. This is a more general case than the traditional one discussed in the previous paragraph. In the literature on optimal control, there's no explicit mention of "multicriteria". However, the optimality criterion involves the state of the system as well as the input control function, and is in general of the form

$$\mathbf{F}(x(t), u(t), t) \rightarrow \min$$

and the different criteria may appear, according to the form of the function $\mathbf{F}$, as combinations of functions of the state vector and the control vector. As references, we can cite [Fel] and [Aok].

Another domain where a weighting of different components is needed is the field of neural networks. A neural network consists of a number of interconnected cells (corresponding to the neurones), and there are weights for the different connections between individual cells. This case is relevant to this discussion since the role of these weights is exactly to emphasise the importance of certain connections over other ones. This is a standard scheme in neural networks. In [AleM], the question of weighting is discussed, and an evaluation and comparison with another approach, the weightless case, is carried out, whereas in [HerKP], the concept of a weight space for a set of neural nets is considered.

Finally, it has been mentioned, in the discussion of the general background to this thesis in chapter one, that one topic, that of estimation of parameters of a given p.d.f. can be considered as a particular case in the scheme of least squares approach. The idea behind this link is that, in such a parameter estimation problem, we have an a priori form for the p.d.f., so that it is not completely specified. Regardless of the origin of this adopted form for the density, expressed in terms of the parameters sought for, it can be considered as some kind of assumptions made on this p.d.f, which is the essence of the least squares method. However, since this type of questions relates to the analytical side of the problem, a more thorough discussion is left to the next section.

## 4.2. Formulation of the problem

To start the analytical formulation of the problem in the least squares sense, we'll repeat, with more thorough discussions, the brief presentation at the beginning of this chapter.

We have a finite sequence of moments $(m_k)_{1 \leq k \leq N}$, from which we propose to construct, or estimate a suitable p.d.f. having this sequence of numbers as its moments up to the order $N$. In fact, it will be seen that, in the general case, we may obtain, in a set of possible alternatives, functions which moments differ from these given ones, and these alternatives will appear to be, in the context of the formal "syntax" of the method, acceptable; this point will be discussed later, when it appears

86

in the form of the solution. Now the moments appear in the Taylor expansion of the characteristic function as the coefficients, and it is natural to think of forming the truncated characteristic function

$$\Phi_T(\lambda) = \sum_{k=0}^{N} \frac{i^k m_k}{k!} \lambda^k \qquad (4.2.1)$$

And since the p.d.f is the inverse Fourier transform of the c.f., we can think of performing this transformation on the truncated function $\Phi_T(\lambda)$. However, it can be seen that this function is a polynomial in the variable $\lambda$, and so it will diverge for $\lambda$ approaching infinity. Such a function has an inverse (or direct) Fourier transform only in the sense of distributions, where this transform will be a combination of the impulse function $\delta$ and its derivatives. To obtain a regular function, which is the purpose here, we have to force $\Phi_T(\lambda)$ to go to zero for large values of $\lambda$, and it is proposed that this will be done by multiplying $\Phi_T(\lambda)$ by a weighting function, decaying fast enough to zero at infinity. Then it will be possible to perform the required transformation on the new function, which is now

$$W(\lambda)\Phi_T(\lambda)$$

Now given the form of $\Phi_T(\lambda)$ as in (4.2.1), and the arbitrary nature of the function $W(\lambda)$, it can be seen that performing the inverse Fourier transform on the weighted function will not necessarily yield a probability density. In fact, the result may not even be a real function. At this stage, the projection is introduced. What we need is to project the resultant function, $W(\lambda)\Phi_T(\lambda)$, on a space of admissible functions, which in our case here are Fourier transforms of probability density functions. A p.d.f. must be positive and integrable over the real line, with integral equal to one:

$$p(x) \geq 0 \quad \forall x$$

$$\int_{-\infty}^{\infty} p(x)dx = 1$$

However, it is difficult to deal analytically with the positiveness condition. Functions of this kind occupy, in the function space, some kind of a half space, similar to the half line $\mathbf{R}^+$. In fact, in functional analysis, the space of positive functions is an example of a **cone**, and optimisation problems corresponding to this kind of subsets pertain to the convex, non-differentiable case. In our case, we don't need these tools, at least at a first stage. It is better to consider the space of all real valued functions, which form a linear space over $\mathbf{R}$. The space of Fourier transforms of such functions is again a linear space over $\mathbf{R}$. So we can project our weighted function on this space, perform the inverse transformation, and check for positiveness. The second condition is a normalisation condition and can be dealt with easily. Now the projection is performed analytically by minimising the square distance of the weighted function to the mentioned space

$$\int_{-\infty}^{\infty} \left| W(\lambda)\Phi_T(\lambda) - \Phi^*(\lambda) \right|^2 d\lambda \to \min \tag{4.2.2}$$

where $\Phi^*(\lambda)$ is the preliminary required solution( the * doesn't mean complex conjugation), that is, it will still need checking for positiveness and normalisation. To have a complete formulation, it would be important to characterise the space of admissible functions, the space from which the preliminary solution $\Phi^*(\lambda)$ is chosen. It has been defined as the space of Fourier transforms of real valued functions. That means that the inverse Fourier transform of $\Phi^*(\lambda)$ must be a real valued function. In that case, we have

$$\overline{\Phi^*(\lambda)} = \Phi^*(-\lambda)$$

an expression which can be derived from the expression of the Fourier transformation

$$\Phi^*(\lambda) = \int p(x)\exp(i\lambda x)dx$$

and remarking that $p(x)$, being real, satisfies $\overline{p(x)} = p(x)$, while we have

$$\overline{\exp(-i\lambda x)} = \exp(i\lambda x)$$

and the result is that the real part of $\Phi^*(\lambda)$ is an even function, while its imaginary part is an odd function, and this applies for every function which is the Fourier transform of a real valued function.

Now, in the so obtained formulation of the problem, there's an apparent discrepancy with the usual form, also presented and discussed here, of a weighted least squares formulation. This discrepancy resides in the placement of the weighting function in the statement of the problem. In the form we adopted, given by (4.2.2)

$$\int \left| W(\lambda)\Phi_T(\lambda) - \Phi^*(\lambda) \right|^2 d\lambda \qquad (4.2.2)$$

whereas in the usual formulation we should have something like

$$\int W(\lambda)\left| \Phi_T(\lambda) - \Phi^*(\lambda) \right|^2 d\lambda$$

It can be seen that in the first form, the weighting function multiplies only one of the two terms in the difference, while in the second, more usual form, the weighting applies to the whole difference. To resolve this discrepancy, remark that the role of the weighting adopted for our specific problem is to force the otherwise divergent $\Phi_T(\lambda)$ to go to zero for $\lambda$ large, and the function $\Phi^*(\lambda)$ is assumed to have this property, and its weight is absorbed in its very existence. Moreover, since another role of the weighting function is to represent the assumptions made in the observation and estimation process, then multiplying $\Phi^*(\lambda)$ with the same weight as $\Phi_T(\lambda)$ would introduce correlation between the two functions, an undesirable effect, since we are seeking the most free solution **after taking assumptions in account**. So the operation of these assumptions is concentrated on $\Phi_T(\lambda)$ in the formulation. The point is that the role of the weighting operation is not only to de-emphasise values of $\Phi_T(\lambda)$ for large values of the variable $\lambda$, but also to weight other values, the overall effect being of giving preference for some points over the others (which is the essence of making assumptions) in a way that may be complicated, and so the effect of multiplying the required solution $\Phi^*(\lambda)$ by the same weighting function may be misleading and confusing.

One kind of possible assumption is to impose a certain form on the solution p.d.f., by making it depend on a certain number of parameters. These parameters may be coefficients in a linear combination of a set of base functions, thus generating a linear space on which we have to project. The combination may be more complicated, may be by allowing the base functions to depend in a non linear way on the parameters. An example for the linear case could be, say,

$$\alpha \exp(-x^2) + \beta \exp(-(x-1)^2)$$

a combination of two Gaussian distributions. For the non-linear case, we may allow the mean and variance of each one of the components to vary, and it's clear that the resultant function depends in a non-linear way on these new parameters. However, all these cases can be covered by the mathematical concept of a differentiable manifold, which is the non-linear extension to the linear space concept. Now the weighting function formulation can deal with such a case by adjusting the weighting function so that the projection, accomplished linearly, falls on the part of the initial, large linear space of transforms of real valued functions consisting in this specific differentiable manifold. These remarks show, as it was anticipated, that the problem of parameter estimations for a parameter dependent p.d.f. may be cast in the context of the least squares formulation. The interpretation of such a situation is that, since we have adopted a general form for the solution, then in some way we have control of this solution, and the resulting weighting function will, as mentioned earlier, visualise the kind of assumptions underlying the adoption of this specific form for the p.d.f. (or $\Phi^*(\lambda)$). The observer will adjust this visualisation according to subjective criteria depending uniquely on himself by means of "tuning" $\Phi^*(\lambda)$. For this specific case, where we have a general form of the p.d.f. in terms of some parameters to estimate, the problem can be described as an adaptive control problem, where $\Phi^*(\lambda)$ is given as a reference input and the system automatically adjusts $W(\lambda)$ to minimise the error. As we have said, here we have $W(\lambda)$ as an output to be visualised and evaluated according to subjective considerations. Thus we can add a loop to represent this operation performed by the observer. We have the following diagram                    below                    in                    Fig.(4.2.1)

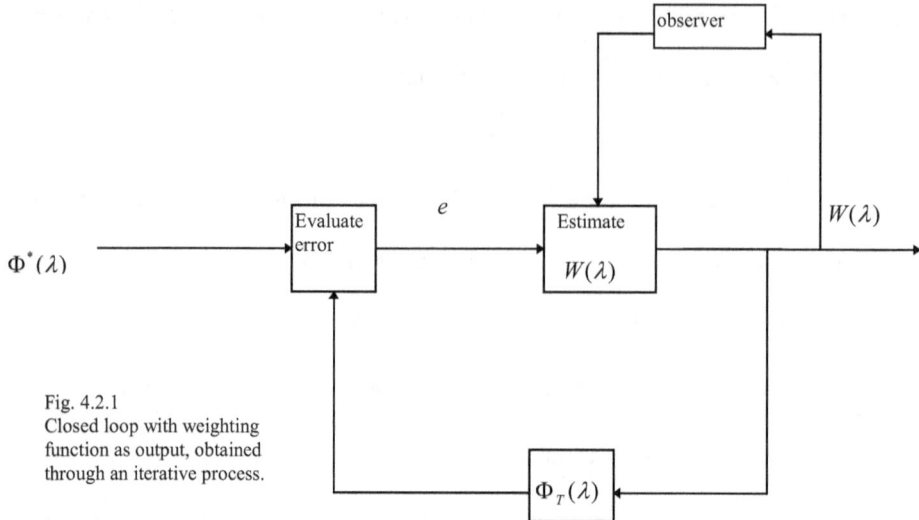

$\Phi^*(\lambda)$

Fig. 4.2.1
Closed loop with weighting
function as output, obtained
through an iterative process.

The error calculated from the difference $W(\lambda)\Phi_T(\lambda) - \Phi^*(\lambda)$ is used, in conjunction with $\Phi^*(\lambda)$, to obtain the weighting function $W(\lambda)$. This latter is fedback through the block $\Phi_T(\lambda)$ and the result is compared to the reference in this case, which is $\Phi^*(\lambda)$. This operation is repeated until the error is minimised and there's a lock-in, or a match, between expectation and reference. To show this, a loop is added to this diagram, the action of which is performed by the observer. According to this diagram, the observer contributes to the lock mechanism by adjusting the weighting function. This is the scheme which would be in action to model the parameter estimation approach in the context of least squares method. On the other hand, the more usual operation, which in some sense dual to this one, can be represented in a diagram where the estimate $\Phi^*(\lambda)$ is the actual output, whereas the input, or control, is the function $W(\lambda)$:

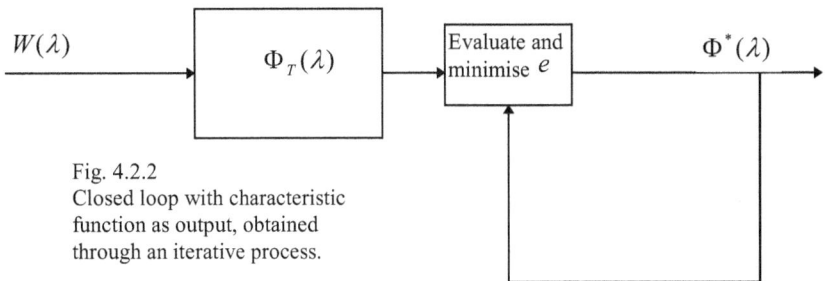

Fig. 4.2.2
Closed loop with characteristic
function as output, obtained
through an iterative process.

Here the updating of the solution is performed automatically by the loop to give the

solution $\Phi^*(\lambda)$ as output. This output shall follow the variations of the input, which

is $W(\lambda)$. This latter will vary through operations of the observer, thus representing

his assumptions and expectations. However, this has not been represented in the loop

diagram because the action of the observer in this case is the ordinary, usual one.

These discussions close the stage of formulation, which has been looked at from

many angles so as to be well clarified. We can proceed now to study the possible

solution, or solutions, of this problem of least squares approach.

### 4.3. Proposed solution

Recall the form of the minimisation problem (4.2.2)

$$\int \left| W(\lambda)\Phi_T(\lambda) - \Phi^*(\lambda) \right|^2 d\lambda \rightarrow \min \tag{4.2.2}$$

This functional will be denoted by $J$. In the context of Hilbert spaces, this problem

can be stated as

$$\|x - a\|^2 \rightarrow \min \tag{4.2.2'}$$

where $a = W(\lambda)\Phi_T(\lambda)$ and $x = \Phi^*(\lambda)$ and where the square norm, represented in

the above formula by $\|.\|^2$ is given in fact by $\int |.|^2 d\lambda$. This is the usual form of the norm in a Hilbert space, especially the space $L^2$ and its variants. The corresponding bilinear form is given formally by the expression

$$\langle x | y \rangle = x \overline{y}$$

and explicitly, for functions

$$\langle f(t) | g(t) \rangle = \int f(t) \overline{g(t)} dt$$

Now the derivative of expression (4.2.2') can be deduced from the derivative of the square norm obtained in chapter 2, section 2.4, where we obtained the linear functional which is the derivative of the square norm function as

$$L.h = \langle x | h \rangle + \langle h | x \rangle \tag{4.3.1}$$

Now the space where our functions are in is the Hilbert space over **R** of square integrable complex valued functions defined on **R**. The subspace on which the projection is taking place is the subspace of real valued functions. We will require a somehow different inner product which makes real valued functions and purely imaginary functions orthogonal. This requirement is dictated by the need for a real valued inner product, and we'll see, after the general solution is obtained the advantage and necessity of such a different scalar product. This will be defined as

$$\langle u | v \rangle_{new} = \frac{1}{2} \{ \langle u | \overline{v} \rangle + \langle \overline{u} | v \rangle \} = \text{Re} \langle u | \overline{v} \rangle$$

so that the scalar product of two functions will be given by

$$\langle f | g \rangle_{new} = \frac{1}{2} \{ \int f(\lambda) \overline{g(\lambda)} d\lambda + \int \overline{f(\lambda)} g(\lambda) d\lambda \} = \text{Re} \int f(\lambda) \overline{g(\lambda)} d\lambda \tag{4.3.2}$$

For one real valued function $f$, and one imaginary valued function $ig$, we'll have

$$\langle f | ig \rangle_{new} = \frac{1}{2} \{ \int f(-ig) d\lambda + \int f(ig) d\lambda \} = 0$$

Moreover, in the dual domain, we'll have that the transform (or inverse transform) of a real valued function and the transform of an imaginary valued function are also orthogonal, since the Fourier transform is an inner product conserving mapping. This can be verified using the characterisation of such functions obtained in the previous section; namely, if we have two functions $f(\lambda)$ and $g(\lambda)$ satisfying the relations

$$f(-\lambda) = \overline{f(\lambda)}$$
$$g(-\lambda) = -\overline{g(\lambda)}$$

and writing their usual inner product

$$\int_{-\infty}^{\infty} f(\lambda)\overline{g(\lambda)}d\lambda = s$$

if we change $\lambda$ into $-\lambda$

$$s = \int_{-\infty}^{\infty} \overline{f(\lambda)}g(\lambda)d\lambda = -\int_{-\infty}^{\infty} \overline{f(\lambda)}g(\lambda)d\lambda = -\overline{s}$$

and it can be seen that this usual innerproduct is a pure imaginary, so that its real part is zero, and subsequently, the new inner product is zero. Now this new product will be denoted $\langle .,. \rangle$. For the case here, we have rather the square norm of a difference to be differentiated, and inserting $x - a$ in the place of $x$, we will have, formally

$$L.h = \langle x - a, h \rangle = \mathrm{Re}\langle x - a | h \rangle$$

In (4.2.2), $x$ is $\Phi^*(\lambda)$ and $a$ is $W(\lambda)\Phi_T(\lambda)$, so that

$$L.h = \mathrm{Re}\int(\Phi^*(\lambda) - W(\lambda)\Phi_T(\lambda))\overline{h(\lambda)}d\lambda \qquad (4.3.3)$$

And we require that this differential to be zero for optimal $\Phi^*(\lambda)$, minimising (4.2.2)

$$L.h = 0 \qquad (4.3.4)$$

Now remark that the choice of the variation $h(\lambda)$ is not completely arbitrary, since it is a variation of $\Phi^*(\lambda)$, which has been subject to restrictions discussed in the previous section, and $h(\lambda)$ will be subject to the same restrictions or constraints, because when $\Phi^*(\lambda)$ moves, it will stay in the subspace prescribed there, and so its variation $h(\lambda)$ must stay in this subspace; then it will satisfy the conditions found in the previous section, namely, if $h(\lambda) = h_1(\lambda) + ih_2(\lambda)$, then

$$h_1(-\lambda) = h_1(\lambda)$$
and
$$h_2(-\lambda) = -h_2(\lambda)$$

(4.3.5)

Let's now write

$$W(\lambda)\Phi_T(\lambda) - \Phi^*(\lambda) = R(\lambda) = R_1(\lambda) + iR_2(\lambda) \qquad (4.3.6)$$

Now the condition of stationarity (4.3.4) implies

$$\text{Re} \int R(\lambda)\overline{h(\lambda)}d\lambda = \int R_1(\lambda)h_1(\lambda)d\lambda + \int R_2(\lambda)h_2(\lambda)d\lambda = 0 \qquad (4.3.7)$$

for all $h_1(\lambda)$ even and $h_2(\lambda)$ odd. In particular, the null function is both even and odd, and by making $h_1(\lambda)$ and subsequently $h_2(\lambda)$ null, we'll have

$$\int R_1(\lambda)h_1(\lambda)d\lambda = \int R_2(\lambda)h_2(\lambda)d\lambda = 0 \qquad (4.3.8)$$

for all $h_1(\lambda)$ even and $h_2(\lambda)$ odd. This can happen only if $R_1$ is odd and $R_2$ is even. This condition for $R(\lambda)$ is the reverse of the one corresponding to the transform of a real valued function. Since the real and imaginary parts will be interchanged if we multiply by the imaginary number $i$, specifically

$$\text{Re}(iR) = -\text{Im}(R)$$
$$\text{Im}(iR) = \text{Re}(R)$$

then $iR$ is the transform of a real valued function, or, equivalently, $R$ is the transform of a function on the real line whose values are pure imaginary numbers, that is, a pure imaginary function. Recall now that we have

$$R(\lambda) = \Phi^*(\lambda) - W(\lambda)\Phi_T(\lambda)$$

and that $\Phi^*(\lambda)$ is the Fourier transform of a real valued function. If we write

$$W(\lambda)\Phi_T(\lambda) = \Phi^*(\lambda) + R(\lambda) \qquad (4.3.9)$$

Where we have changed $-R$ into $R$, an operation which doesn't change the property of this function of being the transform of a pure imaginary function, then we can see that we have decomposed $W\Phi_T$ into two orthogonal components, one, $\Phi^*$, is the transform of a real valued function, the other, $R$, is the transform of an imaginary valued function. We are looking for the density corresponding to $\Phi^*$, and it is clear that it can be obtained by taking the inverse transform of $W\Phi_T$, and then taking the real part. This can be written in the following way

$$p^*(x) = F^{-1}\{\Phi^*\} = \operatorname{Re} F^{-1}\{W\Phi_T\} \qquad (4.3.10)$$

where $F^{-1}$ is the inverse Fourier transform operator. The problem now is the choice of the weighting function $\hat{w}_1(\lambda)$, and more particularly, the choice which will yield a positive p.d.f. $p^*(x)$.

The first step is to remark that the polynomial $W(\lambda)$ in the Fourier domain is in fact a differential operator with constant coefficients in the dual domain; that is, if we write down

$$\Phi_T(\lambda)W(\lambda) = (1 + i\lambda + \cdots + \frac{i^k \lambda^k}{k!} m_k)W(\lambda) \qquad (4.3.11)$$

and putting $w(x) = F^{-1}\{W(\lambda)\}$, then

$$F^{-1}\{\Phi_T(\lambda)W(\lambda)\} = \sum_{k=0}^{n} \frac{(-1)^k m_k}{k!} w^{(k)}(x) \qquad (4.3.12)$$

This can be seen in the following way: Let $f(x)$ and $\hat{f}(\lambda)$ be a Fourier transform pair. Then we have, for the case where $\hat{f}(\lambda)$ is a characteristic function

$$f(x) = \frac{1}{2\pi} \int \hat{f}(\lambda) \exp(-i\lambda x) d\lambda$$

now differentiating both sides with respect to the variable $x$, we obtain

$$f'(x) = \frac{1}{2\pi} \int (-i)\lambda \hat{f}(\lambda) \exp(-i\lambda x) d\lambda$$

so that $-i\lambda \hat{f}(\lambda)$ is the transform of $f'(x)$. Repeating this operation a number of times, we deduce that $(-1)^k i^k \lambda^k \hat{f}(\lambda)$ is the transform of $f^{(k)}(x)$. Applying this to (4.3.11) will give the result (4.3.12). Writing now $w(x) = w_1(x) + iw_2(x)$, the solution given by (4.3.10) is

$$p^*(x) = Dw_1(x) = \sum_{k=0}^{n} \frac{(-1)^k m_k}{k!} w_1^{(k)}(x) \tag{4.3.13}$$

where $D$ is the differentiation operator

$$D = \sum_{k=0}^{n} \left( \frac{(-1)^k m_k}{k!} \frac{d^k}{dx^k} \right)$$

with the convention that for the value $k = 0$, $m_0 = 1$ and $\dfrac{d^0 w}{dx^0} = w$. If we let now $\hat{w}_1(\lambda)$ be the transform of the real part $w_1(x)$ ($\hat{w}_1(\lambda)$ is not the real part of $W(\lambda)$), then we have

$$\Phi_T(\lambda)\hat{w}_1(\lambda) = \Phi^*(\lambda) \tag{4.3.14}$$

and so we can consider, without loss of generality, that the weighting function is the Fourier transform of a real valued function. This is the first step in the search of some conditions or characterisations of a possible weighting function. The next step is to investigate one property of the concept of weighting: The role of weighting is to give some "contributions" to the characteristic function, or to $\Phi_T(\lambda)$, more importance than other contributions. This implicitly contains the fact that we have an order relation in the range of the weighting function, which lies inside the complex numbers field $\mathbf{C}$, and we know that only the field of real numbers has this property. This remark may suggest to consider a real valued weighting function. Now if take this option, together with the fact the $W(\lambda)$ is now assumed to be the transform of a real valued function, we deduce from the relation

$$W(-\lambda) = \overline{W(\lambda)}$$

another relation, based on the fact that the conjugate of a real number is the same number itself

$$W(-\lambda) = W(\lambda) \qquad (4.3.15)$$

and this function is to multiply a polynomial, $\Phi_T(\lambda)$. However, this polynomial has no symmetry property in general, and given the fact that the whole $\Phi_T W$ is now a characteristic function (equation (4.3.14)), and accordingly must satisfy a relation similar to (4.3.15), that is, $\Phi_T(-\lambda)W(-\lambda) = \overline{\Phi_T(\lambda)W(\lambda)}$, we can see that this is unfeasible. Another way to see that, and which provides more insight, is to expand the now real valued $W(\lambda)$ into its Taylor series

$$W(\lambda) = \sum_{l=0}^{\infty} a_l \lambda^l$$

where the coefficients are now real numbers. Multiplying with $\Phi_T(\lambda)$, we obtain an expansion of $\Phi^*$ where the coefficients have non-zero real and imaginary parts, whereas they must be of the form $i^k b_k$. These remarks exclude the possibility of a real valued weighting function, and we must introduce a non-zero phase in the numbers $W(\lambda)$. This operation can be interpreted as "shifting" the position of the problem so as to be well posed, and that it is originally an ill-posed problem. This shifting is performed by a phase shift on $W(\lambda)$, making it a complex number. Note, however, that in general, the moments of the estimated p.d.f. will not match exactly the prescribed moments. This can be seen if we expand $W(\lambda)$ into its Taylor series, obtaining an expansion similar to a characteristic function (since its the transform of a real valued function), and then perform the multiplication of the two terms in (4.3.11), to see that the new coefficients of the Taylor expansion are not necessarily equal to the $m_k$. This point will be studied more thoroughly in forthcoming discussions, and especially after the presentation of the maximum entropy method.

Having settled this point, another important remark in the process of choosing $W(\lambda)$ is to realise that one characteristic of the problem as formulated in this chapter is there will be always some freedom in the choice of this function, but in the same time this choice should not be completely arbitrary. These two contradictory requirements can be satisfied together if we let $W(\lambda)$ to vary in some specified set, described, say, by a certain number of parameters, obeying some functional equations, etc.. And if we want to treat the problem in its full generality, this set must be universal, in the sense that it is defined, once and for all orders of moments considered. Looking at equation (4.3.13), we see that a requirement is that this set must be closed under all orders of differentiation. Roughly said, the set must stay the same after differentiation. In this intuitive language, something similar is satisfied by an exponential function. For reasons to appear more clearly later, and because it satisfies these requirements, we propose that this set be the set of functions of the form

$$f(x) = P(x)\exp(Q(x)) \tag{4.3.16}$$

We can check easily that

$$f'(x) = (P'(x) + P(x)Q'(x))\exp(Q(x))$$

Repeating this operation any number of times and adding the results, as is performed by a linear differential operator with constant coefficients like the one in (4.3.13), the result will be of the same form as (4.3.16). This set will be denoted subsequently by **PE** .As for the problem of non-matching between obtained moments and prescribed ones, mentioned earlier in this section, we can hope to approach these prescribed moments in the case of the above mentioned set by increasing the order of one of the polynomials $Q$ or $P$, or both. The examples in the following section will illustrate this point to some degree.

## 4.4. Extension to higher dimensions

The extension of the results obtained so far to the vector random variable case is straightforward. We have now the multidimensional moments

$$m_{i_1 \cdots i_n} = E(X_1^{i_1} \cdots X_n^{i_n})$$

from which we build the truncated characteristic function

$$\Phi_T(\lambda_1, \ldots, \lambda_n) = \sum \frac{i^{|k|} m_{k_1 \cdots k_n}}{k_1! \cdots k_n!} \lambda_1^{k_1} \cdots \lambda_n^{k_n}$$

where $|k| = k_1 + \cdots + k_n$ and the powers in the above summation are less than a specified value of $|k|$, but not necessarily all of these powers appear in the summation. From here, there's nothing particular about the solution of the problem which differs from the simple monodimensional case. The problem is formulated in the following way

$$\int_{-\infty}^{\infty} \left| \Phi_T(\lambda_1, \ldots, \lambda_n) W(\lambda_1, \ldots, \lambda_n) - \Phi^*(\lambda_1, \ldots, \lambda_n) \right|^2 d\lambda_1 \cdots d\lambda_n \to \min$$

and the solution is obtained , in the $x$-domain by taking the inverse transform of the product $\Phi_T W$ and then taking the real part. Note that in this case, the differential operator corresponding to $\Phi_T$ is of the form

$$\sum \frac{m_{k_1 \cdots k_{k_n}}}{k_1! \cdots k_n!} \frac{\partial^{|k|}}{\partial \lambda_1^{k_1} \cdots \partial \lambda_n^{k_n}}$$

## 4.5. Some specific calculations

In this section, some computations will be carried out to find the candidate solutions for the problem for the first few orders of the moments.

We begin with order zero, where only the normalisation condition is enabled, and there's no restriction on the values of the mean and higher order moments. In that case, the differential operator in (4.3.13) will be of order zero, and is equal to the identity, mapping $w(x)$ (or $w_1(x)$) into itself, and we have the solution

$$p^*(x) = w(x)$$

that is, we can chose any function from the "pool" of functions defined in the previous section. We can say that the strategy implied by the least squares method in the case of complete uncertainty is to make any "possible" choice, from this pool, and to chose the coefficients as to ensure positiveness and normalisation. However, in practice, the effective p.d.f. will be needed over a certain finite interval, say for

$$a \leq x \leq b$$

In such a case, and to preserve the whole real line formulation, we can weight the chosen solution further with another function from the same pool, taking care to preserve as much of the original density as possible. For this operation, we can use the functions

$$f_p(u) = \exp(-u^{2p}) \tag{4.5.1}$$

These functions vanish for $u \to \pm\infty$, since the exponent of $u$ in (4.5.1) is even, and they are flat for values of $u$ around zero, and the extent of "flatness" increases with the exponent $p$. That's because the derivatives at zero of such functions are zero up to the order $2p-1$. Now to implement this operation for the interval considered above, we chose a point inside this interval, say $c$, and we set

$$u = x - c$$

in (4.5.1).A possible point is the middle of the interval. The new weighted function will be almost equal to the original one around $c$, and negligible when the variable $x$ is outside the interval. The new density will have the form

$$p^*(x)\exp(-(x-c)^{2p}) \tag{4.5.2}$$

and this function belongs also to the set of admissible functions adopted previously. The exponent $p$ has to increase with increasing interval length. This approach will be used in subsequent examples of higher order moments, and is applicable to all similar cases.

We turn now to the first non trivial case corresponding to moment order one. Equation (4.3.13) now is

$$p^*(x) = w(x) - m_1 w'(x) \tag{4.5.3}$$

If we chose an arbitrary, generic function from the set **PE**, such as $P(x)\exp(Q(x))$, as the function $w(x)$ what we'll obtain is

$$p^*(x) = (P(x) - m_1 P'(x) - m_1 P(x)Q'(x))\exp(Q(x))$$

However, we can obtain a simple looking solution if we chose a small order polynomial $Q(x)$. A particular choice is suggested by solving the homogeneous differential equation obtained from (4.5.3), that is,

$$w(x) - m_1 w'(x) = 0 \tag{4.5.4}$$

which will yield

$$w(x) = C\exp(\frac{x}{m_1}) \tag{4.5.5}$$

The idea is that since we are looking for a positive p.d.f., then by adding a certain number to the exponent in (4.5.5), we can hope that the left hand side of (4.5.4) will be greater than the right hand side, that is, greater than zero. Now in general, a change in the values of a function on certain points or over a certain interval is uncorrelated to changes in its derivative. This can be seen, for an arbitrary function $f(x)$, by adding a strongly oscillating function $\varepsilon(x)$, with $0 < \varepsilon(x) << |f(x)|$. The function $f(x) + \varepsilon(x)$ is close to $f(x)$, but the derivatives are very different because of the strong oscillation assumption. However, in the case here of an exponential function, changes in the derivative are correlated to changes in the exponent. To see this, consider the function $\exp(\alpha x)$. Its derivative is $\alpha \exp(\alpha x)$. Adding $\varepsilon$ to the exponent, the derivative becomes $(\alpha + \varepsilon)\exp((\alpha + \varepsilon)x) > \alpha \exp(\alpha x)$ for $\alpha, \varepsilon > 0$. We can apply this result to (4.5.4) and (4.5.5). Suppose that $m_1 > 0$, that we are working in an interval where also the variable $x > 0$, and chose $\alpha < \dfrac{1}{m_1}$. Set

$w(x) = C\exp(\alpha x)$, $C$ being for normalisation. Then we have

$$p^*(x) = w(x) - m_1 w'(x) = C(1 - m_1\alpha)\exp(\alpha x) > 0$$

since $\alpha < \dfrac{1}{m_1}$ and the normalisation constant is also positive. If we are considering a finite interval such as $a \le x \le b$, the solution would be

$$p^*(x) = \frac{\exp(\alpha b) - \exp(\alpha a)}{\alpha}(1 - m_1\alpha)\exp(\alpha x) > 0$$

Again, if the solution needs to be expressed over the whole line while staying inside the set $\mathbf{PE}$, we can multiply with a certain function as in (4.5.1) and (4.5.2). If we now compute the first moment of this distribution

$$m_1^* = \int_a^b x p^*(x)dx$$

we obtain a complicated expression which is close to $m_1$ for $\alpha \to \dfrac{1}{m_1}$ :

$$m_1^* = \frac{1}{\alpha}\left[1 + C(\alpha,a,b)(m_1\alpha - 1)\right]$$

$C(\alpha,a,b)$ includes the normalisation constants and numbers arising from the integration. But choosing such a value for $\alpha$ would lead to a solution close to zero. This can be seen from the very argument which led to the solution: Recall that we have the expression of the solution

$$p^*(x) = C(1 - m_1\alpha)\exp(\alpha x)$$

Choosing a value $\alpha \to \dfrac{1}{m_1}$ would make the term $(1 - m_1\alpha) \to 0$. The way out of this difficulty is to use the pool of functions defined previously, that is, the set $\mathbf{PE}$, choosing a higher order polynomial in the exponential part of the candidate function, or a higher order polynomial multiplying the exponential. We can begin with the second alternative, choosing

$$w(x) = (\alpha x + \beta)\exp(\gamma x) \tag{4.5.6}$$

and we assume we are working on a finite interval $[a,b]$. Then the solution will be, according to (4.3.13)

$$p^*(x) = w(x) - m_1 w'(x) = \left[(\alpha x + \beta) - \gamma m_1(\alpha x + \beta) + \alpha\right]\exp(\gamma x) =$$
$$\left[\alpha(1 - \gamma m_1)x + \beta(1 - \gamma m_1) + \alpha\right]\exp(\gamma x)$$

This function has to satisfy the normalisation condition over $[a,b]$. We have three unknowns, $\alpha$, $\beta$, and $\gamma$. The normalisation condition will leave only two of them, the third one, say $\gamma$, is a function of the other two, $\alpha$ and $\beta$ (and certainly $a$ and $b$). The first moment is given by

$$m_1^* = \int_a^b x p^*(x)dx$$

If we want this moment to be equal to the prescribed one, that is, $m_1^* = m_1$ we finish with one equation with two unknowns, $\alpha$ and $\beta$, and in general, we can be sure we can adjust these two variables so as to reach the equality $m_1^* = m_1$. What we have done is to add one degree of freedom in $w(x)$. The same will hold if we chose to use a second order polynomial in the exponential in $w(x)$, although the solution may then be a different one. This procedure can be carried out for higher orders of moments chosen, since we are free to chose from the pool **PE** polynomials and exponents of any order.

# CHAPTER 5

# MAXIMUM ENTROPY METHOD

This chapter presents the final method proposed in this thesis for the estimation of a probability density from its moments, and it will appear in this discussion as the most attractive procedure to accomplish this task. First we begin by a review of the concept of entropy, since this concept has appeared in different contexts before, and the connection of these different variants with the core subject in this thesis will be clarified.

## 5.1. A review of the concept of entropy

Historically, the concept of entropy was introduced first in thermodynamics by Carnot in the nineteenth century, where he stated the existence of a state function which should increase with time, or at least may not decrease with time (It is constant in a reversible transformation of the system). This function is called the entropy of the system, and the previous statement is known as the second law of thermodynamics [Zam]. But it was only when it was introduced in information theory by Shannon [Sha] that its link with probability theory and probability distributions has become clear. At the actual state of knowledge, the entropy of a continuous random variable $X$, with a continuous p.d.f. $p(x)$, is given by

$$H(X) = -\int p(x)\ln(p(x))dx = E(\ln(\frac{1}{p(X)}))$$   (5.1.1)

and for a discrete event setting, that is, a discrete r.v., the entropy is defined as

$$H(X) = -\sum_i p_i \ln p_i .$$   (5.1.2)

In this section, the discussion of the possible different aspects of the entropy of a r.v.

will be carried out on the discrete event case, the results being readily generalisable to the continuous case.

Consider now a set of discrete events $A_i$ having probabilities of occurrence $p_i$. If, for a certain $A_i$, $p_i$ is too small or too close to one, the uncertainty about the occurrence of such an event is low, since it is almost sure that this event will not happen, or will happen, respectively. The quantity $p_i \ln(\frac{1}{p_i})$ is zero at these two extremes, and has a maximum in the interval $0 < p < 1$ for $p = \frac{1}{e}$ (for logarithm base $n$, this would occur for $p = \frac{1}{n}$). These remarks show that the entropy of a r.v. as defined in (5.1.2) is in fact a measure of the uncertainty in the random variable $X$. Now let's consider again a low probability event $A_i$. Then the occurrence of such an event carries much information about the observed process, since it is not expected. This is in contrast with the case of an almost sure event ($p_i$ close to 1), where the occurrence does not carry much additional information. But for every low probability event, there should be one or more high probability events, and again the sum in (5.1.2) represents the amount of information carried by the random variable, but now it is the **average** information content, since there is weighting of $\ln p_i$ (large in magnitude for small $p_i$) with $p_i$ and summation over all the events. This point brings the analysis to the observation of many realisations of the same r.v. or, equivalently, to the observation of a certain number of i.i.d r.vs. where we have a sequence of observed data. And since in this case we have many observations, the problem of estimation of a p.d.f. can be posed, where we're about to infer from the sequence of data a certain probability law, which, according to previous discussions, represents information contained in these data. Following the foregoing remarks on

108

the information carried by an observation, we can ask if the sequence in hand is a "normal" event according to the required p.d.f. or not. If it is not, it is then a rare event which does not represent the actual state of things and is to be discarded. However, the normal decision, **if there are no other external reason** to do so, this sequence is to be kept and considered valid. This would be the normal behaviour and decision. Another way of seeing that is that considering the data as rare events must trigger an interest for searching the reason behind this abnormal behaviour, and this leads to the study of the problem in the light of these actual data. This point settled, the fact that the data observed is representative or typical reminds us of the concept of complexity discussed in the first chapter, where one of the situations was that the data sequence is of typical complexity. These sequences are characterised by the fact that the frequency of occurrence of events in them scales with their probabilities of occurrence. It can be shown [CovT], [Pap] that the number of such sequences (of length $n$) is close to $2^{nH(X)}$, so that the probability of occurrence of a particular one of them is close to $2^{-nH(X)}$, the result being that the occurrence of a sequence of typical complexity is close to unity. A sequence which deviates a little from this typical or normal behaviour will have a very low probability of occurrence [Pap], [CovT]. In the light of all these different aspects of entropy, the maximum entropy principle for estimation of a p.d.f. is to chose one which achieves maximum uncertainty, maximum average information, and such that the data presented to the observer is of typical complexity according to the same required p.d.f. Two more points are needed to clarify the problem. First, for the complexity point of view, the p.d.f. estimated by the maximum entropy method generates the maximum number of sequences of such typical complexity, according to the previously mentioned number of these sequences, where the number of these sequences is an exponential of the entropy of the r.v.. Recalling the second principle of thermodynamics, stating that the entropy of a system must increase with time, and the statistical mechanics interpretation of entropy as the number [Pat]

$$S = k \ln \Omega \qquad (5.1.3)$$

(this equation is due to Boltzmann)where $\Omega$ is the number of "microstates" in a certain macrostate of the system, this latter being characterised by state variables as the energy, temperature, molecular density, etc.., we see the connection between the two situations. In thermodynamics, the system will evolve in time to the macrostate which has the maximum number of microstates underlying it. Any one of these microstates will generate the same state of the system. In a problem of p.d.f. estimation, when presented with a sequence of data, the maximum entropy method will chose the p.d.f. for which there's a maximum number of sequences of this complexity, which is chosen as typical. The second point concerns the role of the moments. This role can also be understood in the light of the complexity paradigm. In practice, we are presented with a certain sequence, whereas the foregoing discussion treated a large number of "similar" possible sequences. Here appears the analytical role of the moments estimated from the data, for these moments, put in the front of the observer, will represent a general pattern of the data, more general than the specific sequence in hand. The moments have a number of different sets of data underlying them, and they give the observer access to all of these different sets or sequences.

It remains that the main characteristic of entropy of a system or r.v. used in this thesis is the fact that it is linked to uncertainty. The maximum entropy state of a system, for example, is one where the uncertainty about what specific microstate it is in is maximal. For the problem of estimating a probability density, the maximum entropy method will chose the p.d.f. which contains just the available information, not more, and this choice again is characterised by maximum uncertainty. Analytically, this can be seen in the concept of mutual information [CovT][Pap]: Given two r.vs $X$ and $Y$ we define the mutual information between them as

$$I(X,Y) = H(X) - H(X|Y) \qquad (5.1.4)$$

Where the second term on the right hand side is the conditional entropy, that is, the entropy of $X$ knowing $Y$. It can be shown that this quantity (the mutual information) satisfies the inequalities

$$0 \leq I(X,Y) \leq H(X) \qquad (5.1.5)$$

That means that the uncertainty about $X$ has been decreased by the amount of conditional entropy, which is the information about $X$ contained in $Y$. Introducing assumptions outside the range of available constraints decreases the entropy, and thus the maximum entropy estimate of a density consumes zero amount of additional information.

## 5.2. Historical and modern background

As mentioned in the previous section, the concept of entropy in science and technology appeared for the first time in thermodynamics and was introduced by Carnot in the nineteenth century. In the late forties (1948)of this century [Sha], entropy was introduced by Shannon in the context of information theory, a theory which he laid the basic foundations. Later, in the sixties, Kolmogorov defined the entropy of an ergodic transformation. Common to all these aspects of the concept of entropy is the connection with the general paradigm of information content. But the context most linked to the theme of this thesis was in the work of Jaynes [Jay] who enunciated the maximum entropy principle for the estimation of a probability law or probability density. This result can be found in the original paper as well as in

[KapK]. It states that for all p.d.fs generating a set of data and subject to certain constraints, almost all of them lie very close to the maximum entropy solution, the distance being the deviation of the expected frequencies of occurrence of events for a generic p.d.f. from the ones given by the maximum entropy distribution [KapK]

$$\Delta = \frac{1}{2N} \Sigma \frac{(Np_i - Nq_i)^2}{Np_i}$$

$N$ is the number of hypothetical samples, the $p_i$ stand for the maximum entropy density, and $q_i$ for a density slightly deviating from the maximum entropy solution. Also in this latter reference, as well as in [Pap], the exponential law for the maximum entropy solution subject to certain constraints of the form

$$E(f_i(X) = g_i, i = 1,\ldots,n$$

is obtained, but no mention of any particular role for the moments. Also, for the sake of completeness, we have to mention the use of the maximum entropy principle in spectral estimation, a method due to Burg. In the one dimensional case, this method is known to give the same solution as the MA modelling of the signal approach [Pri].

## 5.3. Formulation of the problem

To begin the formulation, recall the expression of the entropy of a continuous r.v. given by (5.1.1)

$$H(X) = -\int p(x) \ln(p(x)) dx \qquad (5.1.1)$$

We'll look at the continuous case. Here, the only information available is given by the moments of $X$ up to an order $N$, any additional information being excluded, so that the solution is expected to be one having the prescribed moments with no external assumptions. These moments may be estimated from experimental data, or exact numbers coming, for example, from theoretical densities. We have

$$m_k = \int x^k p(x)dx \, , \; k = 0,1,\ldots, N \qquad (5.3.1)$$

so that the constraints are

$$g_k(x) = \int x^k p(x)dx - m_k = 0 \qquad (5.3.1')$$

We have to maximise $H(p) = H(X)$ subject to these constraints. To do so, we form

the Lagrange function

$$J(p,\lambda_0,\lambda_1,\ldots,\lambda_N) = H(p) - \sum_{k=0}^{N} \lambda_k g_k(x) \to \max \qquad (5.3.2)$$

That is, this functional has to maximised with respect to $p(x)$, the required solution

being the function $p^*(x)$ at which the maximum occurs.

## 5.4. Proposed solution

Let's find the Frechet derivative of $J(p) = J(p,\lambda_0,\ldots,\lambda_N)$. Let $h(x) = \delta p(x)$ be

a small variation of $p(x)$. Then

$$J(p+h) = \underbrace{- \int (p(x)+h(x))\ln\left[p(x)+h(x)\right]dx}_{J_0(p+h)} - \underbrace{\sum_{k=0}^{N} \lambda_k \left[\int x^k (p(x)+h(x)) - m_k\right]}_{J_\lambda(p+h)}$$

(5.4.1)

We have to calculate $J(p+h) - J(p)$, where $J(p)$ is also written in the form

$$J(p) = J_0(p) - J_\lambda(p) = -\int p(x)\ln(p(x))dx - \sum_{k=0}^{N} \lambda_k \left[\int x^k p(x)dx - m_k\right]$$

Using the fact that $h$ is a small variation and that $\ln(1+u) \approx u$ to the first degree, we

can write

$$J_0(p+h) = -\int (p+h)\ln\left[p(1+\frac{h}{p})\right]dx = -\int (p+h)(\ln p + \frac{h}{p})dx$$

to obtain, after developing and discarding terms of higher order than one in $h$

$$J_0(p+h) = -\int p\ln p dx - \int h(1+\ln p)dx$$

then

$$J_0(p+h) - J_0(p) = -\int h(1 + \ln p) dx$$

For the second term

$$J_\lambda(p+h) - J_\lambda(p) = \int (\sum_{k=0}^{N} \lambda_k x^k) h dx$$

The total variation of the Lagrange function with respect to $p(x)$ is now

$$J(p+h) - J(p) = -\int h(1 + \ln p) dx - \int (\sum_{k=0}^{N} \lambda_k x^k) h dx = -\left[ \int (1 + \ln p + \sum_{k=0}^{N} \lambda_k x^k) dx \right].h = J'.h$$

and $J'$ is the required derivative, which should be zero at the extremum:

$$1 + \ln p^*(x) + \sum_{k=0}^{N} \lambda_k x^k = 0$$

hence

$$\ln p^*(x) = -1 - \lambda_0 - \lambda_1 x - \cdots - \lambda_N x^N$$

$$p^*(x) = \exp(\sum_{k=0}^{N} \alpha_k x^k) \qquad\qquad (5.4.2)$$

where $\alpha_0 = -1 - \lambda_0$ and $\alpha_k = -\lambda_k$. So we have obtained a specific function as a solution. This function is positive over the whole line. Now another requirement is that the estimated p.d.f. must go to zero at infinity. For $N$ even, this will be the case if the coefficient $\alpha_N$ is negative, since then $\alpha_{2p} x^{2p} \to -\infty$ for $x \to \pm\infty$ where $N = 2p$ and subsequently $\exp(\sum \alpha_k x^k) \approx \exp(\alpha_{2p} x^{2p}) \to 0$. On the other hand, for $N$ odd, the polynomial in the exponent of (5.4.2) will go to $+\infty$ for $x \to +\infty$ or $x \to -\infty$ depending on the sign of $\alpha_N$. The reason for this behaviour could be the fact that for manipulating an integral like $\int x^k p(x) dx$, it must be absolutely convergent. This is always the case for $k$ even ($|x|^{2p} = x^{2p}$) as long as the required density $p(x)$ satisfies adequate growth conditions for $x \to \pm\infty$. For $k$ odd, if $k+1 \le N$, $\int x^{k+1} p(x) dx$ is absolutely convergent since, for $|x| > 1$ (which is the case because we are considering the behaviour for $x \to \pm\infty$)

$$|x|^k < |x|^{k+1} = x^{k+1}$$

but for $k = N$, this is not necessarily true, because the constraint corresponding to the moment of order $N+1$ is not present. To treat such cases, we can either consider the practical case of a finite interval or use the method suggested in the previous chapter, consisting in using a flat weighting function on the corresponding interval of the form $\exp(-(x-c)^{2p})$. That implies that we can assume, when necessary, that the interval considered is the whole real line, and an even higher order exponent is present in the polynomial of (5.4.2). A final point on the form of the solution (5.4.2), before further discussions, is that this form of solution belongs to the set **PE** defined in the previous chapter, and the result (5.4.2) supports this choice among other reasons, and the connection between the two solutions, least squares and maximum entropy, will appear to be more tight in the analysis carried out in the next chapter.

The coefficients $\alpha_k$ in (5.4.2) are related to the moments of the same solution (5.4.2) by the set of relations

$$m_k = \int x^k \exp(\sum_{l=0}^{N} \alpha_l x^l) dx, k = 0,1,2,... \qquad (5.4.3)$$

from which we deduce two useful sets of relations: First, taking the partial derivatives in (5.4.3) of both sides with respect to the coefficients $\alpha_l$

$$\frac{\partial m_k}{\partial \alpha_l} = \int x^k \frac{\partial}{\partial \alpha_l} \exp(\sum_{i=0}^{N} \alpha_i x^i) dx = \int x^k x^l \exp(\sum_{i=0}^{N} \alpha_i x^i dx) = \int x^{k+l} \exp(Q(x)) dx$$

where $Q(x) = \sum_{i=0}^{N} \alpha_i x^i$. The last integral is nothing but $m_{k+l}$ so that, finally

$$\frac{\partial m_k}{\partial \alpha_l} = m_{k+l} \ , \ 0 \le l \le N, \ k = 1,2,\dots \tag{5.4.4}$$

Now let's integrate (5.4.3) by parts. Then

$$m_k = \left[ \frac{x^{k+1}}{k+1} \exp(Q(x)) \right]_{-\infty}^{\infty} - \int \frac{x^{k+1}}{k+1} Q'(x) \exp(Q(x)) dx \tag{5.4.5}$$

assuming that the integration is carried out over the whole real line, a feasible assumption according to the previous remark on the integration boundaries. The first term in the above expression is zero, while the second is

$$-\frac{1}{k+1} \sum_{p=1}^{N} p \alpha_p m_{k+p} \ \text{so that, finally}$$

$$m_k = -\frac{1}{k+1} \sum_{p=1}^{N} p \alpha_p m_{k+p} \ \ k = 0,1,2,\dots \tag{5.4.6}$$

In this equation, the order $N$ may be the genuine order of the prescribed moments, or the artificial one introduced by using the weighting $\exp(-u^{2p})$. If we are interested in a limited interval only, we have to evaluate the first term in (5.4.5) between the boundaries of this interval. We obtain

$$m_k = \frac{b^{k+1}}{k+1} \exp(Q(b)) - \frac{a^{k+1}}{k+1} \exp(Q(a)) - \frac{1}{k+1} \sum_{p=1}^{N} p \alpha_p m_{k+p}, k = 0,1,2,\dots \tag{5.4.6'}$$

These equations can be used to compute the effective coefficients $(\alpha_k)_{0 \le k \le N}$. Here we will make use of (5.4.4) to build an algorithm to compute these coefficients as will be explained later. Now we can compute the entropy which is supposed to be maximal for the solution (5.4.2) adopted. To do so, recall the expression of the entropy (5.1.1)

$$H(X) = -\int p(x) \ln(p(x)) dx \tag{5.1.1}$$

Using (5.4.2), this can be written as

$$H(X) = -\int (\exp(\sum_{k=0}^{N} \alpha_k x^k))(\sum_{k=0}^{N} \alpha_k x^k)dx =$$

$$-\sum_{k=0}^{N} \alpha_k \int x^k \exp(\sum_{k=0}^{N} \alpha_k x^k)dx = -\sum_{k=0}^{N} \alpha_k \int x^k p(x)dx$$

The last result in these equations is clearly a combination of the moments, so that, finally

$$H(X) = -\sum_{k=0}^{N} \alpha_k m_k \qquad (5.4.7)$$

This result for the value of the entropy was obtained in [KapK] but for the case of general constraints of the form $E(f_i(X)) = g_i$. Let's look now at some specific cases, namely for the first values of the order $N$:

$N = 0$. The corresponding p.d.f. is the exponential of a polynomial of order zero: $p(x) = \exp(\alpha)$. It is a constant over a certain interval $[a,b]$. The strategy implied by the maximum entropy method towards complete uncertainty is to chose a specific distribution, the uniform distribution. This is consistent with the interpretation of the maximum entropy approach as making no additional assumptions: The uniform distribution gives no preference for any point on the interval over any other point, since there is no information which can justify such a preference.

$N = 1$. The solution is of the form $p(x) = A\exp(\alpha x)$ over a finite interval $[a,b]$, or, if we chose to work on the whole line, $p(x) = A\exp(\alpha x)\exp(-(x-c)^{2p}), c \in [a,b]$.

$N = 2$. Then $p(x) = \exp(Q(x))$, where $Q(x) = a + bx + cx^2, c < 0$. This will be always a Gaussian distribution. To show that, let's write

$$a + bx + cx^2 = c\left[(x + \frac{b}{2c})^2 - \frac{b^2}{4c^2}\right] + a =$$

$$c(x + \frac{b}{2c})^2 - \frac{b^2 - 4ac}{4c}$$

$$p(x) = \exp(Q(x)) = \exp(-\frac{b^2 - 4ac}{4c})\exp(c(x + \frac{b}{2c})^2) \qquad (5.4.8)$$

This is clearly a Gaussian density of mean $-\dfrac{b}{2c}$ and variance $\sigma^2 = -\dfrac{1}{2c}$. The first

factor on the left hand side of (5.4.8) still contains the parameter $a$, which should be

adjusted to satisfy the normalisation condition. We deduce from this observation, and

from the relations derived previously in this section between the $m_k$ and the $\alpha_k$, that

the estimated p.d.f obtained by the maximum entropy method is a generalisation of

the Gaussian density, characterised by the fact that all of its higher order moments

are functions of the first two. For the general maximum entropy solution, the higher

order moments $m_p, p > N$ depend on the finite set of coefficients $\alpha_k, 0 \le k \le N$,

which themselves depend on the prescribed first $N$ moments.

## 5.5. Extension to higher dimensions

We look now at the case of a multidimensional, or vector, random variable The

extension of the foregoing results is straightforward. We have a number of moments

of the form

$$m_{i_1 \cdots i_n} = E(x_1^{i_1} \cdots x_n^{i_n}) = \int x_1^{i_1} \cdots x_n^{i_n} p(x_1, \ldots, x_n) dx_1 \cdots dx_n$$

and the entropy function to maximise

$$H(\mathbf{X}) = -\int p(x_1, \ldots, x_n) \ln p(x_1, \ldots, x_n) dx_1 \cdots dx_n$$

The solution is obtained through the same procedure as the monodimensional case

$$p^*(x_1,\ldots,x_n) = \exp(\alpha_{i_1\cdots i_n} x_1^{i_1}\cdots x_n^{i_n})$$

The equivalent of the equations (5.4.4) is now

$$\frac{\partial m_{k_1\cdots k_n}}{\partial \alpha_{l_1\cdots l_n}} = m_{k_1+l_1\cdots k_n+l_n}$$

## 5.6. General constraints (not moments)

As mentioned in section two of this chapter, the maximum entropy method for estimating a p.d.f. from constraints appears in previous literature [Pap],[KapK]. These constraints are of the form

$$E\{f_i(X)\} = g_i, \ i = 1,\ldots, N \tag{5.6.1}$$

Where the $f_i(X)$ are functions of the random variable $X$ which can be of any type, depending on the specific situation, and the case where they may be moments of $X$ is not considered in the previously mentioned references. The resulting p.d.f., estimated according to the maximum entropy principle under the constraints (5.6.1) is similar to the result here corresponding to moments. Namely, we obtain [Pap],[KapK]

$$p(x) = \exp(\sum_{i=0}^{N} \beta_i f_i(x)) \tag{5.6.2}$$

where $f_0(x) \equiv 1$, and $\beta_0 = -1$. We will show here that this method can be phrased in terms of the moments, and so that the moment approach is still general enough to include all possible situations. However, the complete outline of the relationship between the type of constraints (5.6.1) and the moment approach will be left to the next chapter, in which various methods of estimation are compared. In this section, a brief analysis will follow.

The first remark is that, since we are estimating a p.d.f., only the local behaviour of the functions $f_i$ is important, for a density goes to zero at infinity and will be to a certain degree concentrated around a point or several points. This fact implies that it is enough to take into account the Taylor expansions of the functions $f_i$ at the point zero. If some of these functions is concentrated at another point, say $x_0$, then taking the expansion around $x_0$ will do, and finally we will end with some polynomial expansion of each of the $f_i$

$$f_i(x) = \sum_{k=0}^{p} a_{ik} x^k \tag{5.6.3}$$

We will assume that the order of the expansion satisfies $p = N$. Now taking the expectation values

$$E\{f_i(X)\} = \sum_{k=0}^{N} a_{ik} E(X^k)$$

to end up with

$$g_i = \sum_{k=0}^{N} a_{ik} m_k \tag{5.6.4}$$

This is a linear system of $N$ equations with $N$ unknowns, the moments, and solving it, we obtain the new constraints in the form of moments. The conclusion is that constraints expressed in the broad form (5.6.1) can be transformed to constraints on the moments.

## 5.7. Algorithm and simulations

In this section, an algorithm will be described which computes the coefficients of the polynomial in (5.4.2), and some simulations will be carried out to show the behaviour of the proposed solution. This algorithm will be based on the equations (5.4.4), which in facts give the derivative of the moments as functions of the required coefficients $\alpha_i$.

### 5.7.1. Algorithm

The first order Newton method for finding the root of an equation of the form $f(x) = \alpha$ will be used, but, as will be seen, higher order variants of the same method will be necessary when dealing with higher moment order.

Recall equations (5.4.4)

$$\frac{\partial m_k}{\partial \alpha_l} = m_{k+l} , \ 0 \le l \le N , \ k = 1,2,\dots \tag{5.4.4}$$

and construct the matrix $\mathbf{U}$ defined by

$$\mathbf{U}_{ij} = m_{i+j}, 1 \le i, j \le N \tag{5.7.1}$$

This matrix is the Jacobian of the function giving the moments in terms of the $\alpha_i$. To begin the algorithm, chose an arbitrary value for the $\alpha_i$, say the vector $\mathbf{a}$. This choice will yield computed values for the moments given by (5.4.3). The resulting vector of moments is denoted by $\mathbf{m}$, whereas the prescribed moments are given in $\mathbf{m}_0$. These may be either estimated from experimental data, or exact moments coming, for example, from theoretical densities Set the error in the corresponding choice of $\mathbf{a}$ to be

$$\mathbf{dm} = \mathbf{m}_0 - \mathbf{m} \tag{5.7.2}$$

then the error in $\mathbf{a}$, where $\mathbf{a} = [\alpha_1 \ \alpha_2 \cdots \alpha_N]^T$, will be

$$\mathbf{da} = \mathbf{U}^{-1}\mathbf{dm} \tag{5.7.3}$$

then the value of $\mathbf{a}$ is updated to become

$$\mathbf{a} = \mathbf{a} + \mathbf{da} \tag{5.7.4}$$

and the process is repeated from the beginning, that is, a new value for $\mathbf{m}$ is computed, the error calculated, and so on. Note that the vector $\mathbf{a}$ defined above does not contain $\alpha_0$. This value is computed from the normalisation condition at each step of the iteration. The iteration continues until a sufficiently small value of the error on the moments is reached. The natural choice of the error expression, and the one actually used in this algorithm, is

$$\Delta m = \frac{\sum |dm_i|}{\sum |m_{0i}|} \tag{5.7.5}$$

The second order Newton method makes use of the second order derivatives of moments with respect to the $\alpha_i$, and in this case we have a tensor

$$\mathbf{V}_{ijk} = m_{i+j+k} \tag{5.7.6}$$

then we compute first $\mathbf{dm}$ and $\mathbf{da}$ as in (5.7.2) and (5.7.3). Now a new vector is computed, given by

$$\mathbf{v} = \mathbf{V}[\mathbf{da}][\mathbf{da}] \tag{5.7.7}$$

finally, a new value for $\mathbf{da}$ is now

$$\mathbf{da} = \mathbf{da} - \frac{1}{2}\mathbf{U}^{-1}\mathbf{v} \tag{5.7.8}$$

and the value of $\mathbf{a}$ is updated as before

$$\mathbf{a} = \mathbf{a} + \mathbf{da}$$

and so on. We can still use higher order Newton methods. The formulae giving these methods are found in [Col].

## 5.7.2. Simulations and analysis

In this subsection, we present a few simulations, using the algorithm described in the previous subsection, comparing the moments of known random variables with their computed counterparts which are the moments obtained by the algorithm. The moment orders chosen here are three and four. The moment order two has been omitted since it was shown analytically to be a Gaussian density. In the pages following this discussion, simulations for some cases are presented, where we pick a known density, and from its first three or four moments we compute the density using the proposed maximum entropy method through the above described algorithm. It should be stressed that it is not meant here to reach the theoretical density from which the moments were derived, but to estimate a pdf which has these theoretical moments as its first 3 or 4 moments. In tables 5.7.1 through 5.7.12, the theoretical moments and the computed ones are presented for the purpose of comparison. The top row of each table is the maximum entropy solution function obtained. The column "Theoretical moments" gives the moments of the corresponding theoretical density, while the column "Computed moments" gives the moments of the solution. The majority of the results were obtained up to an error of less than 1% according to the relative error measure adopted in the previous subsection in the description of the algorithm. Tables 5.7.1 and 5.7.2 present the results for the uniform density over the interval [-1,1], for orders 3 and 4 respectively. In tables 5.7.3 and 5.7.4, the same theoretical density, uniform over [-1,1], is considered, for orders 3 and 4 respectively, but where the interval of integration has been enlarged to [-1.5,1.5], looking at the uniform density as a function having the value of zero outside [-1,1]. Tables 5.7.5 and 5.7.6 give the results for the Gaussian density of mean 0 and variance 1, for orders 3 and 4 respectively. Next, we move on the $x$ axis in the positive direction nad look in thables 5.7.7 and 5.7.8 to the uniform density over [0,1], for orders 3 and 4, respectively. Again, as we have done for the uniform density over [-1,1], we extend the interval of integration for the uniform density over [0,1] to become [-0.5,1.5], and tables 5.7.9 and 5.7.10 show the results. In tables 5.7.11 and 5.7.12, we move also along the $x$ axis for the Gaussian density and consider the Gussian density of mean 0.5 and variance 1.

| $\exp(-0.7 - 0.11x - 0.022x^2 + 0.26x^3)$ | |
|---|---|
| Theoretical moments | Computed moments |
| $m_{01} = 0$ | $m_1 = 0.015$ |
| $m_{02} = 1/3$ | $m_2 = 0.33$ |
| $m_{03} = 0$ | $m_3 = 0.015$ |

Table 5.7.1: Results for moments derived from uniform density over [-1,1], order 3.

| $\exp(-0.8 - 0.17x + 1.1x^2 + 0.4x^3 - 1.3x^4)$ | |
|---|---|
| Theoretical moments | Computed moments |
| $m_{01} = 0$ | $m_1 = 0.0205$ |
| $m_{02} = 1/3$ | $m_2 = 0.336$ |
| $m_{03} = 0$ | $m_3 = 0.0205$ |
| $m_{04} = 1/5$ | $m_4 = 0.195$ |

Table 5.7.2: Results for moments derived from uniform density over [-1,1], order 4.

| $\exp(-0.42 + 0.017x - 1.32x^2 + 0.015x^3)$ | |
|---|---|
| Theoretical moments | Computed moments |

| $\exp(-0.42 + 0.017x - 1.32x^2 + 0.015x^3)$ | |
|---|---|
| $m_{01} = 0$ | $m_1 = 0.01$ |
| $m_{02} = 1/3$ | $m_2 = 0.34$ |
| $m_{03} = 0$ | $m_3 = 0.01$ |

Table 5.7.3: Results for moments derived from uniform density over [-1,1], extended to [-1.5,1.5], order 3.

| $\exp(-0.95 - 0.05x + 2.75x^2 + 0.14x^3 - 3.6x^4)$ | |
|---|---|
| Theoretical moments | Computed moments |
| $m_{01} = 0$ | $m_1 = 0.011$ |
| $m_{02} = 1/3$ | $m_2 = 0.34$ |
| $m_{03} = 0$ | $m_3 = 0.0106$ |
| $m_{04} = 1/5$ | $m_4 = 0.200$ |

Table 5.7.4: Results for moments derived from uniform density over [-1,1], extended to [-1.5,1.5], order 4.

| $\exp(-0.96 + 0.022x - 0.442x^2 - 0.0044x^3)$ | |
| --- | --- |
| Theoretical moments | Computed moments |
| $m_{01} = 0$ | $m_1 = 0.01$ |
| $m_{02} = 1$ | $m_2 = 1.1$ |
| $m_{03} = 0$ | $m_3 = 0.008$ |

Table 5.7.5: Results for moments derived from Gaussian density of mean 0 and variance 1 over [-3,3], order 3.

| $\exp(-0.92 + 0.022x - 0.55x^2 - 0.004x^3 + 0.026x^4)$ | |
| --- | --- |
| Theoretical moments | Computed moments |
| $m_{01} = 0$ | $m_1 = 0.01$ |
| $m_{02} = 1$ | $m_2 = 1.2$ |
| $m_{03} = 0$ | $m_3 = 0.005$ |
| $m_{04} \cong 3$ | $m_4 = 4.25$ |

Table 5.7.6: Results for moments derived from Gaussian density of mean 0 and variance 1 over [-3,3], order 4.

| $\exp(0.8 - 13x + 35x^2 - 23x^3)$ | |
|---|---|
| Theoretical moments | Computed moments |
| $m_{01} = 1/2$ | $m_1 = 0.87$ |
| $m_{02} = 1/3$ | $m_2 = 0.63$ |
| $m_{03} = 1/4$ | $m_3 = 0.48$ |

Table 5.7.7: Results for moments derived from uniform density over [0,1], order 3.

| $\exp(-1 - 63x + 332x^2 - 520x^3 + 251x^4)$ | |
|---|---|
| Theoretical moments | Computed moments |
| $m_{01} = 1/2$ | $m_1 = 0.42$ |
| $m_{02} = 1/3$ | $m_2 = 0.24$ |
| $m_{03} = 1/4$ | $m_3 = 0.14$ |
| $m_{04} = 1/5$ | $m_4 = 0.085$ |

Table 5.7.8: Results for moments derived from uniform density over [0,1], order 4.

| $\exp(-1.2 + 4.5x - 1.72x^2 - 2.8x^3)$ | |
|---|---|
| Theoretical moments | Computed moments |
| $m_{01} = 1/2$ | $m_1 = 0.48$ |
| $m_{02} = 1/3$ | $m_2 = 0.32$ |
| $m_{03} = 1/4$ | $m_3 = 0.24$ |

Table 5.7.9: Results for moments derived from uniform density over [0,1], extended to [-0.5,1.5], order 3.

| $\exp(-1 + 4x - 3.7x^2 + 3.3x^3 - 3.75x^4)$ | |
|---|---|
| Theoretical moments | Computed moments |
| $m_{01} = 1/2$ | $m_1 = 0.5$ |
| $m_{02} = 1/3$ | $m_2 = 0.34$ |
| $m_{03} = 1/4$ | $m_3 = 0.247$ |
| $m_{04} = 1/5$ | $m_4 = 0.197$ |

Table 5.7.10: Results for moments derived from uniform density over [0,1], extended to [-0.5,1.5], order 4.

| $\exp(-1.8 + 0.2x + 1.4x^2 - 0.74x^3)$ | |
|---|---|
| Theoretical moments | Computed moments |
| $m_{01} = 0.5$ | $m_1 = 0.54$ |
| $m_{02} = 1.25$ | $m_2 = 1.3$ |
| $m_{03} \cong 1.4$ | $m_3 = 1.626$ |

Table 5.7.11: Results for moments derived from Gaussian density of mean 0.5 and variance 1, over [-1,3], order 3.

| $\exp(-2.4 + 2.2x + 2x^2 - 3x^3 + 0.7x^4)$ | |
|---|---|
| Theoretical moments | Computed moments |
| $m_{01} = 0.5$ | $m_1 = 0.49$ |
| $m_{02} = 1.25$ | $m_2 = 1.22$ |
| $m_{03} \cong 1.4$ | $m_3 = 1.4$ |
| $m_{04} \cong 3.75$ | $m_4 = 3.23$ |

Table 5.7.12: Results for moments derived from Gaussian density of mean 0.5 and variance 1 over [-1,3], order 4.

# CHAPTER 6

# COMMENTS

In this chapter, a comparison will be carried out between the various methods presented in previous chapters, and the interrelations between these methods are exhibited. The main three methods presented so far are the empirical distribution function, the least squares and maximum entropy. In addition, there's the parameter estimation approach, which was discussed briefly in chapter 4, where it was shown to be a particular case in the more general least squares approach. We begin by looking at the empirical distribution function (edf) and its relation to the moment approach, proposed in this thesis.

In the edf, the distribution, or density, is estimated directly from observed data. We have seen that a key factor in the estimation process was to decide whether the data is enough numerous or not. This decision affects the choice of the scale parameter $h$. Another ingredient in the estimated density is the smoothing kernel $K(t)$. These two ingredients are chosen almost arbitrarily, and their effect on the estimated solution can be judged only by experimental application of this estimation in concrete examples. The use of moments solves the problem of this arbitrariness. The "degree" to which data set are numerous is concentrated on the moments order chosen: It becomes gradual, and step by step. As for the smoothing kernel $K(t)$, the problem with it is that it applies to data directly, and a changing set of data will change the result, whereas if it applies to moments, it applies to a fixed, steady pattern represented in the moments. This is the case in the least squares method, where a weighting function is applied to the truncated characteristic function constructed from the moments, and in the maximum entropy function, where the weighting is itself the required p.d.f., and it is constructed directly from the moments. Finally, as we have seen in chapter 1, there may be situations where we have access directly to moments, not the data. In such cases, the moments approach is necessary.

Let's look now at the two main proposed approaches here, maximum entropy and least squares. Recall that the least squares solution does not necessarily have its first $N$ moments equal to the prescribed moments. This can be seen directly from the product

$$\Phi_T(\lambda)W(\lambda)$$

when expanding both terms in powers of $\lambda$, and taking account of the fact mentioned in chapter 4, namely that we can assume $W(\lambda)$ to be the Fourier transform of a real valued function on the real line. This will yield

$$(1 + im_1\lambda + \cdots + \frac{i^N m_N}{N!}\lambda^N)(\sum_{k=0}^{\infty}\frac{i^k \mu_k}{k!}\lambda^k)$$

Developing the product, we will have the expansion of the c.f. of the solution, its first $N$ moments, as well as all the higher order moments being clearly functions also of the $\mu_k$. This product will have the first $N$ moments identical with the prescribed ones only when we have the condition

$$\mu_k = 0, 1 \le k \le N \tag{6.1}$$

In the $x$-domain, this will be expressed as

$$\int x^k w(x)dx = 0, 1 \le k \le N \tag{6.1'}$$

such a function must have negative values as soon as $k \ge 2$ if it is required to be continuous, since even powers of $x$ are positive. But in principle, nothing forbids the final outcome to be a positive function, after applying the differentiation implied by the inverse transform of $\Phi_T$. This observation suggests the following construction. Consider the maximum entropy solution

$$p(x) = \exp(Q(x)) \tag{6.2}$$

where $Q(x) = \sum_{k=0}^{N} \alpha_k x^k$. We will assume this function is considered on the whole real line, and therefore the order $N$ is even As discussed previously, this will not limit the generality of the discussion. Then we take the Fourier transform of (6.2) to obtain the ch.f.

$$U(\lambda) = \sum_{n=0}^{\infty} \frac{i^n m_n}{n!} \lambda^n \tag{6.3}$$

Consider now the truncated function

$$U_T(\lambda) = \sum_{n=0}^{N} \frac{i^n m_n}{n!} \lambda^n \tag{6.4}$$

Then we can write (6.3) in the form

$$U(\lambda) = U_T(\lambda)\left[1 + \frac{1}{U_T(\lambda)} \lambda^{N+1} R(\lambda)\right] \tag{6.5}$$

$R(\lambda)$ being the rest of the expansion (6.3) after deleting $U_T(\lambda)$, divided by $\lambda^{N+1}$, which is the first power in the rest of the expansion. Now, under the condition that $U_T(\lambda)$, considered as a polynomial in the **real** variable $\lambda$, has no real roots, the second term in the brackets is a well behaved function, and the lowest power in it is $\lambda^{N+1}$, since $U_T(\lambda)$ has no poles at zero and no poles on the line. The final outcome is that the c.f. (6.3), which corresponds to the maximum entropy solution, can be written as the product of $U_T(\lambda)$ with a weighting function, which is the expression inside the brackets in (6.5). The main conclusion is that, under certain conditions, the maximum entropy solution can be recovered from the least squares method by a suitable choice of the weighting function. Now what is the interpretation of this result. The meaning lies in the role of the weighting function of limiting the interval of validity of the polynomial $\Phi_T(\lambda)$. This polynomial is a Taylor expansion, and thus it is an accurate representation of the whole required c.f. only on a certain interval around zero, and this limitation is implemented by the weighting function, which should go to zero for relatively large values of the variable $\lambda$, that is, for values outside the range of validity of $\Phi_T(\lambda)$. Now a function $W(\lambda)$ satisfying (6.1) has null derivatives at zero up to the order $N$, and thus is flat at zero, the flatness being "to order $N$". It can be seen from (6.1') that there may be many functions satisfying these requirements, and we postulate that the maximum entropy one, constructed as in (6.5), is a special one, and a typical one., and should be characterized only by (6.1) or (6.1').

An equivalent connection between the two methods is seen when we realize that, in general, for the typical maximum entropy solution for a moments order of three or higher, the coefficients $\alpha_k$ can be obtained only through numerical computations. We have seen that their relations with the prescribed moments do not yield these coefficients in closed form in terms of the moments, and so they can be obtained up to a certain error. We will show now, in a formal heuristic argument, that if the error measure on the moments is taken to be a relative error, i.e. of the form $\varepsilon(\mathbf{m}) = \dfrac{|\Delta \mathbf{m}|}{|\mathbf{m}|}$, where $|.|$ is a suitable norm, and $\mathbf{m}$ is the infinite, or finite to some order, moments vector, then the approximate p.d.f., corresponding to the approximate coefficients $\alpha_k$, is in fact a weighting function solution, that is, a least squares solution for some specific weighting function corresponding to the above mentioned error. To see this, let's write, formally,

$$\mathbf{m} = \mathbf{m}_0 (1 + \varepsilon(\mathbf{m})) \tag{6.6}$$

and

$$|\mathbf{m}| = |\mathbf{m}_0|(1 + \varepsilon(\mathbf{m})) \tag{6.6'}$$

we can see a one to one correspondence with the formulation of the least squares method, which we recall here

$$\int \left| W(\lambda) \underset{w}{\Phi_T} (\lambda) - \underset{\mathbf{m}_0}{\Phi^*} (\lambda) \right|^2 d\lambda \tag{6.7}$$

This correspondence is clear if we assume that the norm of the moments vector is obtained by taking the integration of the corresponding polynomial over a certain interval

$$|\mathbf{m}| = \sum \int \left| m_k \lambda^k \right|^2 d\lambda \tag{6.8}$$

and letting the result of (6.7) to be of the form $\varepsilon(\mathbf{m})\mathbf{m}_0$, we have, always heuristically

$$w\mathbf{m}_0 - \mathbf{m}_0 = \varepsilon(\mathbf{m})\mathbf{m}_0$$

the term $w\mathbf{m}_0$ can represent the adjusted moments vector, and letting it be equal to $\mathbf{m}$ would yield the relative error expression (6.6). In terms of the relative error, the term $w$ is given by $w = 1 + \varepsilon(\mathbf{m})$ The norm in (6.8) could be justified by the observation that the component $m_k$ of $\mathbf{m}$ has importance determined by the fact that it is the coefficient of the power $\lambda^k$, and this is taken account of by integrating the polynomial over a certain interval. This heuristic argument strengthens the hypothesis of connection between maximum entropy and least squares method, and gives more support for the adopted formulation of the least squares problem in terms of a weighting function. The conclusion is that the computed maximum entropy solution corresponds to the solution of a least squares problem with a suitable weighting function, and, vice versa, that a computed least squares solution, corresponding to a certain choice of the weighting function from the set **PE**, is an approximate solution for the maximum entropy method in practical case problems, with a certain relative error and over a certain specified interval. We will now proceed to further, more analytical and more rigorous computations clarifying this connection and supporting it. These computations will be based on the set of equations (5.4.6) in chapter 5, which we recall here, assuming an even order of the moments

$$m_k = -\frac{1}{k+1}\sum_{p=1}^{2N} p\alpha_p m_{k+p} , \ k = 0,1,2,... \tag{6.9}$$

Remark first the striking analogy between a function from the set **PE** and the function in (6.5). Both are a product of a polynomial in the relevant variable ( $x$ or $\lambda$ ) with a function, which in the case of (6.5) is a weighting function, having all the first coefficients in the Taylor expansion, up to the order of moments considered, equal to zero, except from the constant, which is the value of the function at zero. Now it has been proposed to use functions of the form $\exp(-u^{2p})$ as weighting functions for a special purpose, and it is clear that these functions have their derivatives at the origin, up to order $2p-1$, equal to zero. This suggests to extend the dual polynomial-weighting function from the $\lambda$ domain to the $x$ domain. Accordingly, for some function in **PE** like $P(x)\exp(Q(x))$, we can interpret the first polynomial, $P(x)$ as the expansion of the solution, which may be well the maximum entropy solution, on a certain interval on the line and around some point $c$ inside this interval, and the second term, $\exp(Q(x))$, is of the form $\exp(-u^{2p})$, where $u = x - c$. In turn, in the $\lambda$ domain, the weighting function for the maximum entropy solution will be, or will be approximated, by a product of the form

$$\Phi_T(\lambda)\exp(z_0 - z\lambda^{-2q}) \tag{6.10}$$

for some integer $q$. The $x$ domain solution will then correspond to a moments order of $2q - 1$, and will be given by

$$(\sum_{k=0}^{2q-1} \frac{(-1)^k m_k}{k!} \frac{d^k}{dx^k}) w_q(x) \tag{6.11}$$

according to (4.3.13) in chapter 4, where $w_q(x)$ is the real part of the inverse Fourier transform of the exponential term in (6.10). Now we will use the equations mentioned at the beginning of this discussion, equations (6.9), to compute the moments corresponding to a function of the form $\exp(\alpha_0 - \alpha_{2N}x^{2N})$, and the $\alpha$'s corresponding to the similar function in the $\lambda$ domain. In the first case, we have

$$\alpha_k = 0, \ 1 \le k \le 2N - 1 \tag{6.12}$$

and equations (6.9) become

$$1 = -\alpha_{2N}m_{2N}$$
$$\vdots$$
$$m_k = -\frac{1}{k+1}\alpha_{2N}m_{2N+k} \tag{6.13}$$
$$\vdots$$

so that, finally, all moments of order $2N$ and higher are determined from the first moments $(m_k)_{0 \le k \le 2N-1}$ and the coefficient $\alpha_{2N}$:

$$m_{2N+k} = -\frac{(k+1)m_k}{\alpha_{2N}} = (k+1)m_{2N}m_k \tag{6.14}$$

For the second case, the exponential in the $\lambda$ domain, where the weighting function under consideration is of the form $\exp(z_0 - z_{2q}\lambda^{-2q})$, we have that

$$m_k = 0, \ 1 \le k \le 2q - 1 \tag{6.15}$$

and by equations (6.9)

$$1 = -\alpha_{2q} m_{2q}$$

$$0 = -\frac{1}{2}(\alpha_{2q-1} m_{2q} + \alpha_{2q} m_{2q+1})$$

$$\vdots$$ (6.16)

$$0 = -\frac{1}{k+1}(\alpha_{2q-k} m_{2q} + \cdots + \alpha_{2q} m_{2q+k})$$

$$\vdots$$

The $\alpha$'s are progressively calculated from these equations, whereas the moments are the higher order derivatives at the origin of the weighting function $\exp(z_0 - z_{2q}\lambda^{-2q})$. For the special case of the zero mean Gaussian density, we have the same shape for the density and the characteristic function

$$g(x) = \frac{1}{\sqrt{2\pi}\sigma}\exp(-\frac{x^2}{2\sigma^2})$$

$$\hat{g}(\lambda) = \exp(-\frac{\sigma^2\lambda^2}{2})$$ (6.17)

and the polynomial term disappears.

One aspect of the preceding discussion is the fact that sometimes we have the density to be estimated on a certain interval. But how to determine this interval? It is here that the role of the empirical distribution function, or more precisely, the data from which the moments were eventually computed, appears. The interval chosen must be an interval that contains all the data, or the most "frequent" items in the data. A systematic way to do this is to evaluate the empirical distribution along the lines presented in the third chapter of the thesis, and then to find the interval on which the integral of the obtained density has the value most close to unity, with a certain allowable difference threshold. In terms of the dual observer-environment, this amounts for the settlement of the question of the framework in which to formulate the problem under consideration, or, so to say, the choice of the "alphabet" in which to code the observations.

# CHAPTER 7

# CONCLUSION

In the preceding chapters, the problem of constructing a probability density function from moments was discussed. The moments approach has been laid on solid foundation by looking at aspects of moments as means for maximum information retrieval from observations, as specifying the degree of details required and as presenting a unified view of the observations through concepts of shape and form. This approach has been shown to derive from the deeper concepts of complexity and duality, and the two main methods of pdf estimation were identified: The least squares method and the maximum entropy method. After reviewing the empirical distribution function and putting it in context with the moments approach, these two methods were discussed and complete solutions were obtained for both of them. The known method of estimating of a pdf by parameter estimation was shown to derive from the least squares method. The solutions obtained for the two proposed methods appeared to be similar, both containing the exponential of a polynomial in their expression in closed form, and a following discussion showed that they are closely related: It was shown that the maximum entropy solution can be recovered from the least squares one by a suitable choice of the weighting function and that the least squares solution is in fact the maximum entropy solution computed approximately up to a certain relative error. Finally, a unified solution was obtained which is amenable to analytical computations, where the set of functions of the form $P(t)\exp(Q(t))$ in both the $x$ domain and the Fourier domain, the $\lambda$ domain, has a major role.

# APPENDIX

# RANDOM NUMBER GENERATION

As another application of the results of this thesis, we propose here an algorithm for generation of random numbers having specified moments. The method relies on the results of chapter 5 to estimate the pdf. Here are the description of the algorithm, the program and the complete listing of the program in MATLAB.

## Description of the algorithm and program

The algorithm used to generate the data estimates first the pdf with the specified moments, and then generates the random numbers by operating on numbers generated uniformly in the interval [0,1] through the function rand in MATLAB. This operation is based on the rejection method, described for example in [Leo]. It will be described later in the description of the program.

The computation of the pdf relies on equations (6.16) in chapter 6 in the thesis. We recall them here

$$
\begin{aligned}
1 &= -\alpha_{2p}\mu_{2p} \\
0 &= -(\alpha_{2p-1}\mu_{2p} + \alpha_{2p}\mu_{2p+1}) \\
&\vdots \\
0 &= -(\alpha_{2p-k}\mu_{2p} + \cdots + \alpha_{2p}\mu_{2p+k}) \\
&\vdots \\
0 &= -(\alpha_{1}\mu_{2p} + \cdots + \alpha_{2p}\mu_{2p+2p-1})
\end{aligned}
\tag{A.1}
$$

where the $\mu_{2p+k}$ are the "moments" (in fact, generalised moments) of the function in the $\lambda$-domain given by

$$
\exp(-A\lambda^{2p})
\tag{A.2}
$$

which appears in the solution characteristic function

$$\Phi(\lambda) = (\sum_{k=0}^{2p-1} \frac{i^k m_k}{k!} \lambda^k) \exp(-A\lambda^{2p}) \tag{A.3}$$

Now, expanding (A.2) will show that the first non zero power of $\lambda$ after $\lambda^{2p}$ is the one corresponding to $\lambda^{4p}$, so that all the $\mu_{2p+k}$, $k = 1,...,2p-1$ are zero, and subsequently, as can be easily seen in (A.1), all the $\alpha_{2p-k}$, $k = 1,...,2p-1$ are also zero, leaving us with the expression corresponding to (A.2) in the $x$-domain

$$\exp(-tx^{2p}) \tag{A.4}$$

From now on, the exponent power will be denoted by $N$ instead of $2p$, assuming that it will be always even. An arrangement in the program to insure this condition for all moments orders eventually required will be explained in the description of the program. Finally, the solution pdf is given by

$$p(x) = (\sum_{k=0}^{N-1} \frac{(-1)^k m_k}{k!} \frac{d^k}{dx^k}) \exp(-tx^N) \tag{A.5}$$

Using the relation in (A.1)

$$1 = -t\mu_N$$

and the expansion of (A.2), we have the following relation between $t$ and $A$

$$t = \frac{1}{AN!} \tag{A.6}$$

In the program, $t$ is called par and $A$ is called alpha.

The next main procedure of the program is the generation of the random numbers from the computed pdf. As mentioned previously, the procedure follows the rejection method. For this to work, we assume that, due to the exponential factor in the expression (A.5), this expression is negligible for some value of the variable $x$ called lim in the program, and is given by

$$\lim = (par)^{-(1/N)} (\log(1/epsx))^{(1/N)} \tag{A.7}$$

where epsx is a (small) threshold: For $|x| \geq \lim$, $\exp(-(par)x^N) \leq epsx$. The generation procedure is as follows:

Generate r=rand uniform in [0,1]. Then $temp1 = 2(\lim)r - \lim$ is uniform in the interval [-lim, lim].

sup is a value equal or greater than the maximum of the pdf inside the interval [-lim, lim]. It could be computed numerically, but in the program it is estimated to be equal to $sup = cst + (par)m(N-1)\lim$. cst is the normalisation constant, and the expression is based on the fact that the derivative of the pdf at zero is $(par)m(N-1)$, and the value of the pdf at zero is equal to cst.

2. generate s=rand uniform in [0,1].

3. Then $temp2 = (sup)s$ is uniform in [0,sup].

4. $u = pe(temp1)$ is the value of the pdf (pe) at temp1. If u>=temp2, chose temp1 to be the required number, otherwise reject this value and repeat the process from step 1 (while loop).

We turn now to the description of the program.

## Description of the program

The complete listing is shown at the end of this subsection. The main part is the M-file master.m. It begins by checking the moments order ord variable. If it is odd: rem(ord,2)==1, then the power in the exponential, N, is set to ord+1. Otherwise (if ord is even), then N=ord+2, and we have to specify some value for the additional moment m(ord+1), a value which is set in the program to be zero. To start running the program, some values have to be entered before typing "master", which are:

num, which is the number of data points to be generated.

ord, the moments order specified.

alpha, the constant in the exponential in the $\lambda$-domain mentioned previously.

The moments vector. It is a column vector containing the specified moments. Note that for ord even, a certain value for the additional moment m(ord+1) must be in this vector. In line 5 of master.m, this value has been set to zero, so if it is required to set this moment to another value, this line could be deleted without problem.

epsx, which is a threshold for which the computed pdf is considered to be small enough.

Having entered these values, type master to start the program. But there's a note on the variable alpha. This variable is variance like, and it controls the number of data points required to reach a match with the moments. For alpha large with respect to the second moment for example, we need a large number of data points if we expect the computed experimental moments of the generated data to match the prescribed ones, and with alpha sufficiently small, we can reach this match with fewer data points.

We return now to the description of the program. After checking for the variable ord as described previously, the variables in the program are set to be global. That's because there are function calls which may use these variables. We explain in the following briefly these variables: a and b are the coefficients of the polynomials in an expression of the form P(x)exp(Q(x)): a corresponds to P and b to Q. They are used in the M-files "derv.m" and "polyderv.m" to calculate the successive derivatives of an expression of the previous form P(x)exp(Q(x)). These derivatives are of the same form, so that the procedure derv is called many times until the order (N-1). The variables da and db are the degrees of the polynomials P and Q, respectively. It can be seen that da is set to zero, because, for the specific case in the program, the first derivation operates on an expression with P equal to a constant: $C\exp(-(par)x^{N})$. As for db, it is equal to N. m is the moments vector. e and f correspond to intermediary polynomials in polyderv and derv. The variable chk is inside the while loop in the M-file gen.m, and counts the number of rejections. It is useful only for checking. The variable u is equal to the value of the pdf at the would be number to generate. The variable st is inside the M-file normsn.m, which computes the normalisation constant cst. The other variables have been explained previously.

The program then computes par by the formula (A.6), and sets the coefficient b(1)=-par, which is the coefficient in the exponential. The other coefficients in b are zero. The assignment a=[1] sets the initial value of the vector a, whereas the normalisation constant cst is introduced else where, multiplying the result of the derivation in pe.m. The procedure polyderv is called in the line polyderv. Now the generation begins, by calling gen inside the loop

```
for i=1:num
```

and the data are stored in the array data.

The operation of gen.m has been described previously. It is the procedure which generates the required numbers based on the uniform random number generator rand.

We describe now the other mentioned call in master.m, which is the file polyderv.m. In this procedure, the results of the derivations

$\frac{(-1)^k m_k}{k!} \frac{d^k}{dx^k} (\exp(-(par)x^N))$, which are all of the form $P(x)\exp(-(par)x^N)$, are

stored in the matrix A, of size $(N-1,(N-1)^2+1)$. Since the sizes in the successive

derivations vary, and vectors and arrays must have the same size to be added

together, the maximum size has been chosen for A, and we append zeros to the result

of each derivation (in the expression e=[zeros...] in polyderv). Certainly, derv is

called at each step in the derivation. In this latter M-file, derv.m, the command

polyder is used to take the derivative of a polynomial, and conv for the

multiplication. derv is based on the fact that in an expression $P(x)\exp(Q(x))$, the

derivative is $(P'+PQ')\exp(Q)$. For that, we have the instruction a=polyder(a),

which corresponds to $P'$, and the instructions c=polyder(b) and

f=conv(a,c), corresponding to $PQ'$. To adjust the sizes of these two terms, we

append zeros to a=polyder(a) in a=[zeros...].

The other components are straightforward. The function peraw.m computes the

pdf without the constant of normalisation, using the results of polyderv.m, and the

file normsn.m computes this constant which is cst, then passes it to the function

pe.m which gives the final value of the pdf, that is, normalised.

One detail to be mentioned is the function factr, which is the factorial of a

number. It has been written for the purpose of this program, but there's a built-in

function in MATLAB, which is the gamma function which can be used instead.

Trials of the program for some examples of moments suggest the need for a

relatively large number of data to obtain a match with these specified moments.

Further, Gaussian-like moments seem to impose the Gaussian pattern, again, unless a

larger number is generated. That is, the data generated have the one or two final

moments correspond more to a Gaussian density, for a limited number of data points.

## Listing

The main file is master.m, which calls the other procedures.

### master.m

```
if rem(ord,2)==1
   N=ord+1;
else
   N=ord+2
   m(ord+1)=0;
end
global a b da db N par lim e f m ord
global cst temp1 temp2 u chk st epsx
da=0;
par=1/(alpha*factr(N));
b(1)=-par;
for h = 2:db+1
   b(h)=0;
end
a=[1];
polyderv;
data=zeros(1,num);
for i= 1:num
   gen
   data(i)=number;
end
```

### gen.m

```
temp1=0;
lim=((par).^(-1/N)*((log(1/epsx)).^(1/N));
normsn;
u=cst;
temp2=1+u;
chk=0;
sup=cst+par.*m(N-1).*lim;
while temp2 > u
   chk=chk+1;
   r=rand;
   temp1=2.*lim.*r-lim;
   s=rand;
   temp2=sup.*s;
   u=pe(temp1);
end
number=temp1;
```

### polyderv.m

```
A=zeros(N-1,((N-1).^2)+1);
for p = 1:db-1
   derv;
   e=((-1).^p).*(m(p)/factr(p)).*a;
   e=[zeros(1,(N-1).*(N-1-p)) e];
   A(p,: )=e;
```

end

### derv.m

```
c=polyder(b);
f=conv(a,c);
a=polyder(a);
if (da >= 1)
  a=[zeros(1,db) a];
elseif (da==0)
  a=zeros(1,db);
end
a=a+f;
da=da+db-1;
```

### pe.m

```
function y= pe(x)
global a b da db N par lim e f m ord
global cst temp1 temp2 u chk st epsx
normsn;
y=cst.*peraw(x);
```

### peraw.m

```
function y = peraw(x)
global a b da db N par lim e f m ord
global cst temp1 temp2 u chk st epsx
z=exp(polyval(b,x));
for k = 1:db-1
  z=z+(polyval(A(k,: ),x)).*exp(polyval(b,x));
end
y=z;
```

### normsn.m

```
global a b da db N par lim e f m ord
global cst temp1 temp2 u chk st epsx
st=quad('peraw',-lim,lim);
cst=1/st;
```

# REFERENCES

[Ack]: J. L. Ackrill, Editor: *A new Aristotle reader*. Oxford, Clarendon Press, 1987.

[Ahi]: N. I. Ahiezer & M. Krein: *Some questions in the theory of moments.* American mathematical society, 1962.

[AleM]: I. Aleksander & H. Morton: *An introduction to neural computing.* International Thompson Computer Press, 1995.

[Aok]: M. Aoki: *Optimization of stochastic systems.* Academic Press, 1967.

[Arb]: M. A. Arbib: *Theories of abstract automata.* Prentice Hall, 1969.

[AstW]: K. J. Astrom & B. Wittenmark: *Adaptive control.* Addison Wesley, 1995.

[BarC]: A. R. Barron & Th. Cover: Minimum complexity density estimation. *IEEE Transactions on Information Theory*, IT-37, July 1991.

[BerCR]: C. Berg, J. P. R. Christensen, P. Ressel: *Harmonic analysis on semigroups.* NY Springer-Verlag, 1984.

[BerP]: J. Berstel & D. Perrin: *Theory of codes.* Academic Press, 1985.

[CheL]: Ta-Pei Cheng & Ling-Fong Li: *Gauge theory of elementary particle physics.* Oxford: Clarendon, 1984 (Oxford science Publications).

[Col]: L. Collatz: *Functional analysis and numerical mathematics.* Academic Press, 1966.

[CovT]: Th. M. Cover & J. A. Thomas: *Elements od information theory.* John Wiley, 1991.

[DeuS]: J. D. Deuschel & D. W. Stroock: *Large deviations.* Academic Press, 1989.

[Doob]: J. L. Doob: *Stochastic processes.* Wiley, 1953.

[DymM]: H. Dym & H. P. McKean: *Fourier series and integrals.* Academic Press, 1972.

[Dyn]: E. B. Dynkin: *Markov processes.* Springer-Verlag, 1965.

[EysK]: M. W. Eysenck & M. T. Keane: *Cognitive psychology: A student's handbook.* Lawrence Erlbaum Associates, 1995.

[Fel]: A. A. Feldbaum: *Optimal control systems.* Academic Press, 1965.

[Gau]: C. F. Gauss: *Theory of the motion of the heavenly bodies.* Dover, NY, 1963 (English translation).

[GiaM]: G. B. Giannakis & J. M. Mendel: Identification of non minimum phase systems using higher order ststistics. *IEEE Transactions on Accoustics, Speech and Signal Processing.* Vol 37, 1989.

[Gne]: B. V. Gnedenko: *The theory of probability.* NY. Chelsea Publishing Company, 1967.

[Gra]: R. M. Gray: *Probability, random processes and ergodic properties.* Springer-Verlag, 1988.

[HerKP]: J. Hertz, A. Krogh, R. Palmer: *Introduction to the theory of neural computation.* Addison Wesley, 1991.

[HewR]: E. Hewett & K. Ross: *Abstract harmonic analysis.* Vol 2, Springer-Verlag, 1979.

[IofT]: A. D. Ioffe & A. D. Tihomirov: *Theory of extremal problems.* North Holland, 1979.

[ItzZ]: C. Itzykson & J. P. Zuber: *Quantum field theory.* McGraw-Hill, 1980.

[Jar]: A. El-Jaroudi, T. Akgul, M. Simaan: Application of higher order spectra to multiscale deconvolution of sensor array signals. *IEEE Conference on Acoustics, Speech and Signal Processing,* Vol 4, pp413-416, 1994.

[Jay]: E. T. Jaynes: Information theory and statistical mechanics. *Physical Reviews, 106,* 1957.

[Kai1]: Th. Kailath (editor): *linear least squares estimation.* Dowden, Hutchinson & Ross, 1977.

[Kai2]: Th. Kailath: A view of three decades of linear filtering theory. *IEEE Transactions on Information Theory.* IT-20(2), 1974.

[KalB]: R. E. Kalman & R. S. Bucy: New results in linear filtering and prediction theory. *J. Basic Eng.*, March 1961.

[KapK]: J. N. Kapur & H. K. Kesavan: *Entropy optimization principles with applications.* Academic Press, 1992.

[Kar]: K. Karhunen: Über lineare methoden in der wahrscheinlichkeit rechnung. *Ann. Academia Scientiarum Fennicae.* Series A1, Vol 37, 1947.

[Kol1]: A. N. Kolmogorov: Logical basis for information theory and probability theory. *IEEE Transactions on Information Theory.* IT-14, 1968.

[Kol2]: A. N. Kolmogorov: Interpolation and extrapolation of stationary random sequences. Reproduced in *Linear least squares estimation* by Thomas Kailath, editor.Dowden, Hutchinson & Ross, 1977.

[Kol3]: A. N. Kolmogorov: Stationary sequences in Hilbert space. Reproduced in *Linear least squares estimation* by Thomas Kailath, editor. Dowden, Hutchinson & Ross, 1977.

[Leg]: A. M. Legendre: Méthode des moindres carrés pour trouver le milieu le plus probable entre les résultats de différentes observations. *Mem. Inst. France.*1810.

[LeiR]: F. Th. Leighton & R. L. Rivest: Estimating a probability density using finite memory. *IEEE Transactions on Information Theory.* IT-32, NO 6, Nov 1986.

[Leo]: A. Leon-Garcia: *Probability and random processes for electrical engineering.* Addison-Wesley, 1994.

[Lev]: Levine: *Fondements théoriques de la radiotechnique statistique.* Editions Mir, Moscou.

[Lin]: P. A. Lindsay: *Introduction to quantum mechanics for electrical engineers.* McGraw-Hill, 1967.

[Lind]: W. C. Lindsey: *Synchronization systems in communications and control.* Prentice-Hall, 1972.

[LinM]: D. Lind & B. Marcus: *Symbolic dynamics and coding.* Cambridge University Press, 1995.

[Loe]: M. Loève: *Probability theory.* Van Nostrand, 1963.

[Mas1]: E. Masry: Recursive probability density estimation for weakly dependent stationary processes. *IEEE Transactions on Information Theory.* IT-32, March 1986.

[Mas2]: E. Masry: Multivariate probability density deconvolution for stationary random processes. *IEEE Transactions on Information Theory*, IT-37, July 1991.

[Mcg]: T. P. McGarty: *Stochastic systems and state estimation.* J.W., 1974.

[Men]: J. M. Mendel: Tutorial on higher order statitics (spectra) in signal processing and system theory. Theoretical results and some applications. *Proceedings of IEEE.* Vol 79, NO 3, March 1991.

[Min]: M. Minsky: The society theory of thinking. In *Artificial intelligence. An MIT perspective*, by P. H. Winston & R. H. Brown editors. MIT Press, 1980.

[MinP]: M. Minsky & S. Papert: *Perceptrons: An introduction to computational geometry.* MIT Press, 1988.

[Mis]: R. Von Mises: *Mathematical theory of probability and statistics.* Academic Press, 1964.

[NikP]: C. L. Nikias & A. P. Petropulu: *Higher-order spectra analysis.* Prentice-Hall, 1993.

[Pap]: A. Papoulis: *Probabilty, random variables and stochastic processes.* McGraw-Hill, 1991.

[Par]: E. Parzen: On estimation of probability density function and mode. *Ann. Math. Stat.*, Vol 33, 1962.

[Pat]: R. K. Pathria: *Statistical mechanics.* Butterworth-Heinemann, 1996.

[Pen1]: R. Penrose: *The emperor's new mind.* London: Vintage, 1990.

[Pen2]: R. Penrose: *Shadows of the mind.* Oxford University Press, 1994.

[Pet]: K. Petersen: *Ergodic theory.* Cambridge University Press, 1983.

[Petr]: A. P. Petropulu: Blind deconvolution of non linear random signals. *IEEE Signal Processing workshop on HOS*, 1993.

[Pri]: M. B. Priestly; *Spectral analysis and time series.* Academic Press, 1981.

[PugS]: V. S. Pugachev & I. N. Sinitsyn: *Stochastic differential systems.* J. W., 1987.

[Rao]: M. M. Rao: *Probability theory with applications.* Academic Press, 1994.

[Sha]: C. E. Shannon: A mathematical theory of communication. *Bell System tech. J.*, Vol 27, 1948.

[SohT]: J. A. Sohat & J. D. Tamarkin: *The problem of moments.* American Mathematical Society, 1970.

[Sor]: H. W. Sorenson: *Parameter estimation.* Marcel Dekker, NY, 1980.

[StuO]: A. Stuart & J. K. Ord: *Advanced theory of statistics.* Edward Arnold, 1991.

[Tre]: H. L. Van Trees: *Detection, estimation and modulation theory.* Part 1. J. W., 1968.

[Ver]: M. D. Vernon: *Perception through experience.* London, Methuen & Co, 1970.

[Wie]: N. Wiener: *Extrapolation, interpolation and smoothing of stationary time series, with engineering applications.*NY, Technology Press and Wiley, 1949.

[Yos]: K. Yosida: *Functional analysis.* Springer-Verlag, 1980.

[Zem]: M. W. Zemansky: *Heat and thermodynamics.* McGraw-Hill, 1968.